GEOGRAPHY AND PLANNING

About the Editor

Dr. K.V.Sundaram, Geographer and Planner, started his career as a lecturer at Alagappah University and received his D.Litt. degree from University of Mysore in 1977. He was awarded the prestigious Senior Nuffield Fellowship from UK (1978-79) for his post-doctoral work in the Centre for South-Asian Studies from Cambridge University.

His technical expertise covers Urban & Regional Planning, Economic and Human Resource Development. He worked with FAO, ESCAP, ICIMOD, CIRDAP, UNCRD and UNICEF in many Asian and African countries and with the Indian Planning Commission. He was also the resource person to World Bank, AIT (Bangkok), Karnataka Planning Board and India's 10th Plan. He has published 23 books and 220 research papers and was recently honoured by his friends and colleagues by a book entitled Development Concerns in 21st Century.

During his career he received a Life Time Achievement Award by Centenarian Trust (Chennai) and Bhoogal Ratna Award by NAGI. He was also felicitated at Banaras Hindu University, Centre for Earth Sciences (Thiruvananthapuram), and University of Allahabad and also by the Mayor of Philippines (1973).

As the Founder Chairman of Bhoovigyan Vikas Foundation, he worked towards developing Earth Care Policies in Third World Countries. At the age of 83 he still dreams of establishing a National Care Policy Institute.

GEOGRAPHY AND PLANNING

Essays in Honour of Prof. V.L.S. Prakasa Rao

Edited by

K.V. SUNDARAM

CONCEPT PUBLISHING COMPANY PVT. LTD.
NEW DELHI-110059

ISBN: 978-81-7022-048-0

Edition 2022

Published and Printed by

Concept
Since 1974

Concept Publishing Company Pvt. Ltd.
A/15-16, Commercial Block, Mohan Garden,
New Delhi-110059 (India)
T: 91 11 41101460
E: publishing@conceptpub.com, *W*: www.conceptpub.com

Cataloging in Publication Data--Courtesy: D.K. Agencies (P) Ltd. <docinfo@dkagencies.com>

Geography and planning : essays in honour of Prof. V.L.S. Prakasa Rao / edited by K.V. Sundaram.

pages cm

Reprint. Originally published: 1985.

ISBN 9788170220480

1. Regional planning--India. 2. India--Economic conditions. 3. Agricultural geography--India. 4. Rural development--India. I. Prakasa Rao, V. L. S., 1917- honouree. II. Sundaram, K. V., editor.

LCC HT395.I5G46 2022 | DDC 338.954 23

PREFACE

THIS book is an attempt to put together what its contributors have deemed to be a fitting tribute to Professor Prakasa Rao a doyen among Indian geographers who has rendered more than 30 years of dedicated service to Indian geography. Several of his students, colleagues and friends expressed a desire that a *festschrift* volume should be brought out in his honour and requested me to take the initiative. Accordingly, I wrote to many of his close associates in June 1982 and received a very encouraging response from all of them. Despite this active interest however, this volume has been nearly three years in the making.

The volume contains 19 papers (including a brief biographical sketch of Professor Prakasa Rao), contributed by 22 geographers. Four among these articles (Learmonth, Ramachandran, Nangia and Sundaram) are in the nature of a critical assessment of Professor Prakasa Rao's contributions to Indian geography and planning. The papers by Sundaram, Davies, Stang and Nagel, Chapman and Harriss and Bhat deal with spatial/regional problems. Farmer, Sita and Phadke discuss the central place theory. Kulkarni and Miller and Wanmali deal with service centres. Some aspects of spontaneous and planned agricultural settlements in south-east Asian countries are discussed in the paper by Uhlig. Parvati's paper presents a market-centre locational model while Tewari's paper is based on intra-metropolitan activity location analysis in Bangalore city. The Misras consider the problem of ruralisation in Indian cities. In his paper on "The Unexplored Realm", Deshpande emphasises that geographers should change their point of view and begin studying man-environment relationships/from the point of view of man that is, aspects of social geography.

A wide range of issues in geography and planning have been discussed. It is hoped that the papers will have a great appeal to geographers and planners. Our grateful acknowledgements are due to the Concept Publishing Company who have agreed to publish this volume and also to make a lumpsum provision towards the royalty of this book which is proposed to be used to institute an award in the name of Prof. Prakasa Rao.

K.V. SUNDARAM

V.L.S. PRAKASA RAO
A Select Bibliography

"Area Planning Approach : Integrated Services," Through Select Centres, Department of Finance and Planning, Government of Andhra Pradesh, 1975. (with Waheeduddin Khan and Sudhir Wanmali).

"Bhangiri : A Case Study of a Market Town," Census of Indian Monograph, 12, Some Aspects of Urbanization in South India, Registrar General, Ministry of Health, New Delhi, 1961.

"Central Place Theory" in L.K. Sen (Ed.), *Readings in Micro-level Planning and Rural Growth Centres*, NICD, Hyderabad, 1972.

"Central Place Theory in Retrospect and Prospect," *Annals NAGI*, 1(2), 1981.

"Census Monograph on some Aspects of Urbanization in South India" (in Press).

"District Planning Officers' Training Programme," Working Paper No. 2, Institute for Social and Economic Change, Bangalore, 1973. (with H.R. Mazumdar, G. Thimmaiah and S. Seshaiah).

Development Strategy for an Agricultural Region : A Case Study of Muzaffarnagar District, Institute of Development Studies, University of Mysore, 1976.

"Economic Atlas of Andhra Pradesh", Vishakhapatnam, 1949, (Mimeo) (with V.K. Ramanadhan).

An Exercise in Re-grouping Administrative Units : A Study of Karnataka, 1980 (Mimeo) (with V.K. Tewari).

"The Geographer and the Localization of Industries," *Journal of Madras Geographical Association*, Vol. 17, 1942.

"Geographical Factors and the Ship Building Industry of India", *Calcutta Geographical Review*, Vol. 3, 1941.

Growth Poles and Growth Centres for Regional Economic Development in India, Geneva, 1970. (with R.P. Misra and K.V. Sundaram)

"Growth Promoting and Growth Retarding Factors in Development: Muzaffarnagar District : A Case Study", Research Programmes Committee (Planning Commission) Report, New Delhi, 1969.

Indian Cities : Ecological Perspectives, Concept Publishing Company, New Delhi, 1985 (with V.K. Tewari and Jay A. Weinstein)

"Industrialization and Urbanization in Mysore State", *Applied Economic Papers*, 102, 1961. (with A.T.A. Learmonth).

"Land Use Survey in India : Its Scope and Some Problems", *Proceedings of the International Geographical Seminar*, Aligarh, 1956.

"Land Use Survey in India : Its Scope and Some Problems", *The Oriental Geographer*, No. 2, 1957.

"Macro-Urban Analysis : Geographers' Contribution", *Bombay Geographical Magazine*, Vol. 10(1), 1962.

Migration and Resettlement : Rural Urban Policies, 4, Social Welfare and Development Centre for Asia and the Pacific, Manila, 1980.

"Mobility and Choice of Central Places", *Analytical Geography*, 1(1). Delhi University, Delhi, 1971, (with R. Ramachandran)

"A Note on Soil Classification", *Bombay Geographical Magazine*, No. 4, 1956.

"Planning for an Agricultural Region", with R.P. Misra, K.V. Sundaram (Eds), *Regional Development Planning in India: A New Strategy*, Vikas, New Delhi, 1974.

"Planning the Metropolis : Trends in Southern Cities", *The Hindu*, Madras, 1981. (with V.K. Bawa)

Planning Regions in Mysore State, Indian Statistical Institute, Calcutta, 1960. (with L.S. Bhat)

"The Problems of Metropolitan Region : Geographer's Point of View", *Journal of the Institute of Town Planners*, 1961.

"Readjustment of District Boundaries in Karnataka", Mimeographed, Centre for Human Geography and Ecology,

Institute for Social Economic Change, Bangalore, 1974. (with V.K. Tewari)

"Regional Aspects of Small and Medium Sized Towns of Telangana", Research Programmes Committee, Planning Commission Report, Hyderabad, 1964.

"A Regional Framework for Resource Development", Paper Contributed to the Seminar of the Council of Geographers (India), Cuttack, 1962. (with L.S. Bhat)

"A Regional Framework for Resource Development", *Bombay Geographical Magazine*, No. 1, 1964. (with L.S. Bhat)

"Regional Groupings of the Districts of Madras State", *Indian Geographical Journal*, 1952.

Regional Planning : *Indian Finance*, Calcutta, 1949.

Regional Planning, Vikas, Bombay, 1963.

Regional Planning, Indian Statistical Institute and Asia Publishing House, Calcutta, 1964.

"Regional Planning in the Mysore State : The Need for Readjustment of District Boundaries", Indian Statistical Institute (Mimeo), 1960. (with L.S. Bhat)

Regional Planning : *Principles and Case Studies*, Asia Publishing House, 1962.

"Regional Planning Units for Rearrangement of Administrative Boundaries", *Calcutta Geographical Review*, No. 11, 1974.

Regional Development Planning in India : *A New Strategy*, Vikas Publishing House Pvt. Ltd., Delhi, 1974. (with R.P. Misra and K.V. Sundaram).

"Regional Disparities in India", Paper to the Colloquium on Regional Inequalities of Development, IGU Commission Meetings, Rio de Janeiro, 1971.

Regional Synthesis of Mysore State, Asia Publishing House, 1963. (with L. S. Bhat)

"Regions in Karnataka", Mimeographed, Institute for Social and Economic Change, Bangalore, 1974. (with V.K. Tewari)

"The Role of Geographers in Land use Planning", *Bombay Geographical Magazine*, 1959. (with L. S. Bhat)

The Structure of an Indian Metropolis : *A Study of Bangalore*, Allied Publishers, New Delhi, 1979. (with V.K. Tewari)

Survey of Research in Geography (Ed.), Indian Council of Social Science Research, New Delhi, 1972.

A Survey of Towns in Mysore State, Asia Publishing House, 1964.

"Socio-Economic Study of Vijayawada City", Department of Geography, Delhi University, Delhi (Mimeo), 1973 and Indian Council of Social Science Research, New Delhi, 1975. (with R. Ramachandran)

"Some Aspects of Regional Structure and Development Policy in Andhra Pradesh : A Preliminary Study", *Analytical Geography*, No. 1, 1971. (with R. Ramachandran)

"Some Aspects of Urbanization in South India", *Census of India*, Vol. 1, Monograph No 12, 1961.

"Theoretical Framework for Regional Planning : A Brief Review Note", in R.P. Misra (Ed), *Regional Planning : Concepts, Techniques, Policies and Case Studies*, Prasarang, The University of Mysore, 1969.

Towns of Mysore State, Indian Statistical Institute, Calcutta, 1964.

"Town Planning", *Indian Finance*, Calcutta, 1949.

"Urban Geography", *Indian Geographical Journal*, 1957.

Urban Population Densities : Models and Correlates, Bangalore, Institute for Social and Economic Change, 1977. (with V.K. Tewari).

"Urban Rural Duality in the Regional Structure of Andhra Pradesh : A Challenge to Regional Planning and Development", *Geographische Zeitschrift*, Frantz Steiner Verlag GMBH, Weisbaden, 1962, 1968. (with B.J.L. Berry)

"Urban Rural India", A Paper submitted to the Second Indo-British Seminar in Geography, Cambridge, 1975. (with V.K. Tewari)

"Urban System and the Regional Economy", *Travaux et Documents de Geographie Tropicale*, CEGET No. 45, 1981. (with V.K. Tewari and H. Ramachandran)

"Urban Telangana", *Ekistics*, Vol. 21, 1966.

Urbanization in India, Concept, New Delhi, 1983.

Urbanization in India : Spatial Dimensions, Institute for Social and Economic Change, Bangalore, 1978. (with V.K. Tewari)

Urbanization in Karnataka, Asia Publishing House.

LIST OF CONTRIBUTORS

L. S. Bhat
Professor, Regional Planning, Indian Statistical Institute. S.J.S. Sansanwal Marg, New Delhi, India.

Graham P. Chapman
Department of Geography, University of Cambridge, England.

H.R.J. Davies
Senior Lecturer in Geography, University College, Swansea, U.K.

C.D. Deshpande
Refiner Professor, Currently Editor, *Annals of the National Association of Geographers*, India.

B.H. Farmer
St. John's College, Cambridge England.

Barbara Harriss
Department of Human Nutrition, London School of Hygiene and Tropical Medicine, London, U.K.

Gopal S. Kulkarni
Indiana University of Pennsylvania, Indiana, Pa. U.S.A.

Andrew Learmonth
Professor Emeritus, The Open University, Milton Keynes, U.K.

Vincent P. Miller
Indiana University of Pennsylvania, Indiana, Pa. U.S.A.

H.N. Misra
Geography Department, Allahabad University and the International Institute for Development Research, Allahabad, India.

R.P. Misra
Formerly Deputy Director, United Nations Centre for Regional Development, Nagoya, Japan, Currently Vice-Chancellor, Allahabad University, India.

Peter Nagel
Department of Geography, Technical University, Aachen, West Germany.

Sudesh Nangia
Centre for Regional Development Studies, Jawaharlal Nehru University, Delhi, India.

C. Parvati
Principal, Women's College, Dindigul, Tamil Nadu, India.

V.S. Phadke
Department of Geography, University of Bombay, Bombay, India.

R. Ramachandran
Professor of Geography, University of Delhi, Delhi, India.

K. Sita
Department of Geography, University of Bombay, Bombay, India.

Friedrich Stang
Professor of Geography, Technical University, Aachen, West Germany.

K.V. Sundaram
Joint Adviser, Planning Commission, Govt. of India, New Delhi.

Vinod K. Tewari
Indian Institute of Management, Bangalore.

Harald Uhlig
Professor of Geography, University of Giessen, West Germany.

Sudhir Wanmali
Research Fellow, International Food Policy Research Institute, Washington, D.C. 200363, U.S.A.

CONTENTS

LIST OF TABLES

LIST OF FIGURES

SECTION ONE

1

PRAKASA RAO : A PROFILE

VADHIPARTI LOVA SURYA PRAKASA RAO was born in 1917 at Vishakapatnam in coastal Andhra Pradesh. His early education was in his native town though he later moved to Calcutta for his post-graduation. His thesis on "Land Use Patterns and Problems" based on field studies in the Lower Godavari Region prepared under Prof. S.P. Chatterjee won for him a doctorate in geography from Calcutta University. The years between 1946 and 1957 saw him as Lecturer at City College and later at Calcutta University. It was during his tenure as Senior Lecturer and then Reader at Madras University during 1952-58 that he edited the prestigious *Indian Geographical Journal*—the first such journal in geography which had earlier been associated with such illustrious names as Prof. N. Subramanyam and Dr. George Kuriyan.

In the year 1956, Prakasa Rao was again at Calcutta, this time in collaboration with Prof. A.T.A. Learmonth at the invitation of Prof. Mahalanobis of the Indian Statistical Institute. The ISI studies illustrated the usefulness of regional surveys to Indian planning. Rao's association with ISI continued until 1962. This was an extremely fruitful period in Prakasa Rao's life. His first major contribution at the time was *Towns of Mysore State.* In 1958 he joined the newly-created Regional Planning Unit of ISI as its head and moved with the Unit to Delhi.

In 1962 Prakasa Rao moved to Hyderabad as Professor of Geography in Osmania University. It was during his tenure at Osmania that he established a collaboration with the geography department at Strasbourg University, France. During 1963-64, Rao also directed a project on "Small and Medium Towns of Telangana" sponsored by the Planning Commission. A visiting professorship at the invitation of the University of Chicago between December 1964 and May 1965 also gave him an opportunity to visit and lecture at a number of universities in the United States as well as in Japan. Not long after, in 1966, Prakasa Rao was invited to join the University of Delhi and he continued in this prestigious appointment until 1972. The years 1970-1972 saw him as Director of the Delhi School of Economics and Dean, Faculty of Social Sciences, both posts which he held with great distinction. The period between 1968 and 1972 was a very fruitful one in his life. It was a time when Rao was able to lay firm foundations both for teaching and for advanced research in geography in the University of Delhi. He played an important role in the 21st International Geographical Congress held in New Delhi in 1968 and was Joint Organiser of the International Symposium on Regional Planning. He was also a Member of the Commission on Quantitative Methods in Geography and of the Commission on Regional Aspects of Economic Development, both Commissions under the International Geographical Union. Another notable achievement during this eventful period was the establishment of a collaboration with British geographers and the organisation of the first Indo-British Seminar in Delhi in 1973 which has since become a recurring feature. Four such seminars have been held so far. Prakasa Rao was the Indian team leader for the 1975 Seminar organised at Cambridge. During his tenure of office in Delhi University, Prof. Rao directed three research projects. One of them was a terrain evaluation study in Bikaner and Jaisalmer districts sponsored by the Ministry of Defence, Government of India during 1967-69. The second one was a study on "Growth Promoting and Growth Retarding Factors in Muzaffarnagar District" sponsored by the Planning Commission during 1967-69. The third was a Socio-Economic Survey of Vijayawada City sponsored by the Indian Council of Social Science Research during 1971-73. This was a study undertaken at the instance of

Dr. K.L. Rao who was then the Minister for Irrigation in the Central Cabinet.

Rao's visits to France and Germany in 1967 and 1968 on invitations from these countries afforded him excellent opportunities to study the regional planning efforts made in these countries. As consultant to United Nations ESCAP in October 1969 he continued his work in this area by assisting them in a project on drawing up guidelines for regional planning. A couple of years later in 1971, Rao spent two months as visiting scholar at Syracuse University.

It would be no easy task to enumerate the number of working groups, panels and committees Professor Rao has been a part of, but perhaps a few such could be mentioned. Among these were : Working Group on Planning Regions of India (1970-71), Task Force on Multi-level Planning (1972), ICSSR Working Group on Social Policy Research (1971-72) and Standing Committee on Human, Political and Economic Geography. His associations with the Government of Andhra Pradesh have been close and long-standing as Member, High Power Development Committee, Telangana, expert panel member to advise on medium term planning and as Member of a Technical Committee on the Identification of Backward Areas (1978-81).

In 1972, Prakasa Rao left Delhi to become a Senior Fellow in the newly started Institute for Social Economic Change (ISEC) at Bangalore. He remained in this Institute till 1979. During this period, he directed the socio-economic survey of Bangalore City and its region. This was a monumental work which was published in 1979 under the title *The Structure of an Indian Metropolis* (Allied Publishers, 1979). During his association with ISEC, Dr. Rao also spent a two-month term in the Australian National University as Visiting Professor at the invitation of the Vice-Chancellors Committee.

During 1979-82, Prakasa Rao became an ICSSR Senior Fellow and was attached to the Indian Institute of Management, Bangalore. During this period, he completed a work on *Urbanization in India—Spatial Dimensions* which has since been published by Concept (New Delhi, 1983). This work represents the quintessence of his thought on urbanization and incorporates the findings of his research on various urban issues conducted over a period of 25 years of active professional work.

Prakasa Rao has been associated as Senior Fellow in the Centre for Economic and Social Studies, Hyderabad, a new institution that has been set up with the support of the Government of Andhra Pradesh and continues to offer his best counsel on several research projects that this Institute has taken up.

During the period of more than 35 years of dedicated service to the profession of geography, Professor Rao has made signal contributions and opened up many research frontiers in geography. He has taught, inspired, counselled and trained a whole generation of students in the universities of Madras, Hyderabad and Delhi. He has been a pioneer in many respects. His contributions in the field of urban and regional research in particular have made him an expert in these areas. At a time when regional planning was a vague idea and little talked about, Rao raised a lone voice and vigorously advocated a regional approach to national problems. He himself undertook the challenge of urban and regional research and demonstrated the usefulness of regional surveys through the valuable contributions that he made when he was in the Indian Statistical Institute. In this process, several geographical techniques were researched upon and refined. He also led a band of geographical researchers in the use of quantitative techniques, but he never committed the mistake of substituting the techniques for judgment. He was forever exploring and advocating the use of sophisticated tools for improving the quality of a judgement rather than for making it. His commitment to geography and his insistence on high standards of scholarship were commendable though, at times, they also made him a controversial academician.

Prakasa Rao's home was in many ways an extension of his geography department. Professional colleagues and research scholars working with him were constant visitors at his home. Discussions often spilled over into his leisure hours and the gracious and welcoming hospitality of Mrs. Rao—an exemplar of the traditional *gurupatni*—proved an added inducement for scholars visiting to learn at the feet of their master.

Prakasa Rao's home was indeed a *gurukula* in the true sense of the word considering the number of research scholars who often stayed with him as *sishyas*. Mrs. Rao, a pious and understanding lady, played an admirably complementary role to that of her academician husband and always provided a great moral support to him. It was open house virtually all the time and many were the friends and colleagues who partook of Mrs. Rao's lavish and traditional hospitality. Her unfortunate demise in the year 1982 proved to be a great blow to Prakasa Rao who survived the shock mainly thanks to the support of friends like Waheeduddin Khan at the Centre for Economic and Social Studies.

Prakasa Rao has still a young and alert mind. His frail exterior conceals a mind that has lost nothing of its keenness or agility. In paying him a tribute in this *festschrift*, his admirers hope that he will continue to be a guiding star to his students for many years to come. It is likely that in a brief biographical sketch of this kind many details have been omitted. The Editor hopes, however, that the essentials depicted will far outnumber them in presenting a true to life picture of the man, the academician and the teacher.

2

PRAKASA RAO AS I KNOW HIM

ANDREW LEARMONTH

NEED I say that it is both pleasure and honour to be asked to write about one of my oldest and closest geographer friends ? The pleasure arises from our friendship; the honour from my deep respect for his intellectual stature and for his massive contribution to geography both in general and in relation to India. My assessment of his standing is quite simple that among Indian geographers of his generation he is one of the very best in original ideas, with an extraordinary capacity for hard work carried through to publication, of an order unfortunately not always associated with the man of ideas. This judgment, I must add, is in the context of an Indian geography coming of age in the years when Prakasa Rao's work attained full and independent maturity, years when it became clear that the efforts of the handful of pioneers of a slightly earlier generation were bearing fruit: Indian geographers, while drawing on the scholarship of other countries—in a strictly non-aligned way ! —were evolving a geography to suit and to serve India. And in doing this, they were almost incidentally and fortuitously establishing the international reputation of Indian geography. Internal worries were there, and some remain—about standards of teaching and research, of publication too, and the lack at least until recently of a journal of truly international reputation is surely related to the problem of balkanisation in Indian geography, of diversity and, worse, divergence of interests whether regional, linguistic, factional or of personality cult. Despite all these and other problems, however, the last generation has seen an impact new in its scale of Indian geographers on the international scene. India's best researchers, men and women, clearly rank with the best anywhere in the world—and in this Prakasa Rao has played a major role, with confidence increasing with the years, and yet modestly, even humbly. (Is it too much to suggest that his scholarly humility combines the best traditions of your culture and of ours ?)

I first met Prakasa Rao in 1956. I was seconded from the University of Liverpool to the Indian Statistical Institute (ISI) under the Colombo Plan by arrangement between the governments of India and of the U.K., and in fact was on my way from the headquarters of the ISI at Baranagar to begin two years of work in the then Mysore State, now Karnataka, at that time on the eve of almost doubling in size and population with the reorganisation of the states. I was looking for what in these technical exchange programmes was termed a 'counterpart', an Indian colleague to work on geography and planning with me so that we might learn from each other, and who might head any continuing geographical team in the ISI or for that matter in South India if things should fall out in that way. Prakasa Rao struck me immediately as a potential colleague, clearly a very able man by any standards, imaginative and sensitive, full of ideas, perhaps a little frustrated that heavy burdens of teaching and administration had prevented him from implementing many projects that he carried in his mind's eye—like many a good scholar and conscientious teacher of the time. He had confidence in himself, and yet a great deal of diffidence, even nervousness of manner at times—characteristics that have matured into the scholar's humility already suggested, perhaps, as time has brought fulfilment to replace frustration. I think it was a little later on a visit to Bangalore that Professor George Kuriyan suggested Prakasa Rao as a possible 'counterpart' and if need be successor to me in the ISI's experiments in regional planning, though understandably Kuriyan was at the same time reluctant to lose Prakasa Rao as his anchorman in the Department of Geography in Madras and in the production of the *Indian Geographical Journal.* It took quite a long time for this suggestion to mature ; in the meantime the ISI had sent a promising young mathematical statistician, Manindra Nath Pal, to join the regional survey team in Bangalore (now a full Professor in the ISI), L.S. Bhat had spent a long vacation of survey work, and a varied cross-discipline team was at work on various aspects of the Mysore project. (L.S. Bhat, also a full Professor in the ISI, was later to succeed Prakasa Rao as head of the ISI's regional planning group at their New Delhi campus). It was 1958 before Prakasa Rao was able to join the ISI full-time, though he had already done a great deal of work for the project during vacations etc. At the comparatively late stage of my own work on the Mysore project, at which Prakasa Rao was free to join the ISI, it seemed best to ask him to take charge of one major sector of our work, and he began the research eventually published as *The Towns of Mysore State.* (I should add that the ISI had already commissioned a study of Bangalore by Professor R.L. Singh of Benares Hindu University). I narrate the details about a project of many years ago because the chances of dates of joining and stages of the work did have an influence on Prakasa Rao's very substantial later contributions to geographical scholarship. While his years with the ISI stimulated important writing from him on regional analysis and regional planning, he would I think agree that *The Towns of Mysore State* was at least a main source of the later stream of important works on urban geography and on the Indian city. Other and later chance—or apparently chance—factors have perhaps also influenced the outcomes. Before moving on from the Mysore project of 1956-58, I would like to add a personal note of assessment of it, partly with some modest pride, partly with very real regret. I think Professor P.C. Mahalanobis' initiative in borrowing me from the University of Liverpool did spark off a good deal of good fresh work, some at the time and much more later—by Professors Bhat and Pal and their colleagues, students and associates as well as by the man honoured by this paper and this *festschrift.* I have always since those years regretted that despite Professor Pal's patient instruction of me as his backward pupil we failed to get further into quantitative methods and into what we should now call the borders between geography and regional science, methods and approaches which, ironically in retrospect, were fast evolving during the very years when we were in the field in Karnataka. This was partly because most of my preparatory work before leaving Liverpool was on a quite different, much more limited, project on land and settlement potential in Assam : then as tragically in 1983, Assam was unsettled enough for Professor Mahalanobis to advise strongly against the Assam project on the very eve of my leaving for India, and to persuade me to turn rather to 'fresh woods and pastures new' in the south. Then Walter Isard's stimulating *Location and Space-Economy* arrived in Bangalore just when the whole team was deeply immersed in processing survey data, and I recall distinctly putting the book back in a library almirah as too theoretical by half ! And lastly it was only later that I found out that Prakasa Rao had already anticipated the quantitative revolution with a joint paper with an old college friend on the methodology of composite index formation of groups of regionally differentiated characteristics—racial ones as it happened, with no less a co-author than one of India's greatest mathematical statisticians, C.R. Rao, F.R.S. But for these chances, I believe, the Mysore project might have been methodologically more innovative than it was ; at least during the two years of the actual Colombo Plan project; its real harvest, then, was in later years.

Prakasa Rao took over the ISI's continuing interdisciplinary regional planning team when I left India in 1958, and continued in charge of that work for some four fruitful years—fruitful, yes and yet they might well have been more so. During those years and since, as sole author or with colleagues, Prakasa Rao contributed a great deal to regional thinking; in part his role was to make known to Indian colleagues, especially geographers, some important trends in regional planning in the many overseas countries he visited from 1959 onwards, and partly to a group of geographers and geographer-planners groping for truly indigenous pattern of regional planning, far more appropriate to India's needs than any imported ideas could ever be. Yet I have suggested that these four years might have been more fruitful, and I firmly believe this to be so. That geography and regional planning came to the ISI was due to the personal hunch of that great thinker and great patriot, P.C. Mahalanobis, and his willingness to take the risk that asking a foreigner to spend a couple of years would have a catalytic effect and in fact produce a team of co-workers trusting each other sufficiently to put cooperative research before individual kudos. Prakasa Rao was of course well endowed with the stature and generosity in giving credit where it is due, and was clearly an excellent leader of his team. However, equally important in the ISI hierarchy during these four years was Pitambar Pant, Joint Secretary of the ISI for some years before his tragically premature death. Unless someone close to Shri Pant writes a biography or substantial memoir about him, I am sure that even now it must be difficult for younger Indians to appreciate how important a figure he was as a 'back-room boy' in the early Five-Year Plan periods. It may seem a trivial example of his contributions, that it was almost through a personal crusade and the exercise of his enormous will-power and persuasive talents that India adopted decimalisation so soon and so wholeheartedly after independence ; as he saw, this early adoption, before mechanical and electronic calculators, not to mention computers, should become widespread, must have saved the country countless forms and many many crores of rupees in the decades since then and to come. He also was a great man in quite a different way from Mahalanobis. The tides of change move when they must without regard to individual personalities—and yet the apparently chance juxtapositions of this or that strong personality can and often does affect the exact course of events locally. Pitambar Pant, almost fanatically pragmatic in his country's interests, was juxtaposed with Prakasa Rao the ideas man, sifting concepts and trends in regional planning from home and abroad in the search for the ones most appropriate to India's evolving and changing needs. There seems no doubt that a clash of two personalities, both admirable men, occurred during those years and that this as much as nostalgia for the groves of academia was responsible for Prakasa Rao's return to university and research posts until and indeed after his recent retirement. Thus while as I have suggested Prakasa Rao's contributions to regional planning are substantial, they would surely have been vastly greater but for the chance juxtaposition I have described. However, the change in direction of Prakasa Rao's career in 1962, back to universities and research institutes, also confirmed the chance allocation of urban studies in the Mysore project of a few years before, and led to a stream of studies of Indian cities both massive in data collection and processing and profound in scholarship.

This is not the place for a systematic review of Prakasa Rao's work on urban geography, but it does seem relevant to outline some of its main milestones, and perhaps to attempt in at least an interim way—for one hopes for much more publications in the coming years—to place it in perspective. Even the titles in sequence are suggestive: *A Survey of Towns in Mysore State* (1964); *Urban-Rural Duality in the Regional Structure of Andhra Pradesh* (co-author, 1968); *Some Aspects of Urbanization in South Asia* (1974); *Urbanization in India: A Spatial Dimension* (co-author, 1978); *The Structure of an Indian Metropolis: A Study of Bangalore* (co-author, 1979); *Indian Cities: Ecological Perspectives* (co-author, in Press.); *Urbanization in India: Spatial Dimensions* (1983.) There is a continuity and a wholeness in these contributions to knowledge, and this despite the fact that quite a few of the works are written jointly with others. And we who worked together in Karnataka in 1956-58 can see roots there. Inspired by the National Sample Survey, but aiming to put more geography into sample design, we all gained experience in deploying and training survey teams, in using sample frames with some spatial dimension with confidence that worthwhile spatial trends might be identified and yet with realism concerning the spatial gaps and the limitations in 'multiplying up' from our sample. Except in teaching I have probably followed up this experience less than any other of the principal researchers of 1956-58 – simply because of the paths and preoccupations of my subsequent career. Certainly Prakasa Rao's work since 1958 makes rich and effective use of that experience, duly refined and tempered. Thus his period in the University of Delhi saw his work with R. Ramachandran on landuse and settlement patterns in the Ganga flood plain, and his Osmania phase the survey of small and medium towns of Telangana, and then a real landmark, the team project of socio-economic survey in Vijayawada. There one can see a major step forward in city survey, where sampling and questionnaire survey leads to a picture of the human dynamics underlying more mechanistic 'ecological' surveys using administrative or census areas. This leads on to the ICSSR/Government of Karnataka survey of Bangalore and its region applying these techniques to a much larger and very dynamic metropolis—a complement but much more than a complement to Professor R.L. Singh's survey almost 20 years earlier during the Mysore project of 1956-58. Prakasa Rao's book with Dr. V.K. Tewari on *The Structure of an Indian Metropolis* is remarkable for its presentation equally of the broad patterns of urban development in Bangalore and of the way in which individual men and women, and individual families, the individual molecules as it were of metropolitan dynamism, move and work and live, decide and locate, increase or die. There is a clear link back through the Vijayawada survey to his first sample surveys of Chitradurga or Mandya or Chikballapur, and forward too to his all-India publications.

India and urban geography alike are fortunate that Prakasa Rao has felt able to project his enormous experience over the last 25 years to his two latest books on the all-India urban scene: *Indian Cities* and *Urbanization in India*, the former more ecological, the latter more spatial. Ideas and originality, stimulus and challenge are there a-plenty, yet it is characteristic that Prakasa Rao and Tewari chose as an initial slogan for their joint work a saying from Wolfgang Stolper: '. . . truth must come before originality'.

Prakasa Rao himself would not like it if an essay on him by (I suppose) the oldest of his many foreign geographer friends did not include some critical note. Looking back on his massive contribution to geography, in regional planning but especially in urban studies, I have the feeling that the various books and papers mean more to those of us with a wide knowledge of India than they do to a regional planner or urban geographer who does not know India, or even to a young Indian worker who has as yet not been able to see more than a limited part of such a large and diverse country as India. If I am right, I think the reason lies in the austerity of Prakasa Rao's scholarship. He is a very warm, sensitive and affectionate man, of broad culture—he is very fond of classical Indian music, for instance—as all will agree who have shared the splendid hospitality of Prakasa Rao, his late wife—a most gracious hostess always, and his lively and talented children. He is as capable as any geographer I know of breathing life, colour and vigour into his writing, indeed of raising geographical prose to the level of literature along the lines of the pleas made by Professors J.W. Watson and D.W. Meinig at the 1983 Golden Jubilee meeting of the Institute of British Geographers. I suspect that he is very inhibited about writing descriptive passages, and probably because as a young man he saw so much that was *merely* descriptive (and seldom written at the level of literature to be cherished as such !) that in reaction, even revulsion, he adopted the very austere standards of scholarly presentation that I have referred to and admire—even envy ! Thus one needs to have been in a Bangalore slum and know its warmth and colour, its hospitality and decency, as well as the appalling squalor and affronts to human dignity to which its people are subject, before the moving molecules of the sample surveys quite come to life as one reads Prakasa Rao and Tewari. Please, gentle reader of my tribute to an old friend, do not misunderstand; I am as grateful for Prakasa Rao's scholarship and austerity of standards as I was when I first got to know him in 1957. It is just that, like Oliver Twist, I am asking for more ! Prakasa Rao is still at the height of his intellectual powers, and these are of a high order by anybody's standards. He is still responding to fresh challenges and new projects. With so much solid work, so many major contributions to knowledge, already achieved, perhaps he can write in a slightly more relaxed way without becoming flaccid at all—allow a little more description, and of vigour and colour from elegance to anger according to the topic, to enter his geographical writing. Again please do not misunderstand—he must write what he must and as he must: and we shall all be beholden to him for his next publications. May they be many and successful ! And rewarding in every sense !

3

PRAKASA RAO : The Rise of Indian Geography

R. RAMACHANDRAN

IT IS perhaps imprudent to attempt an assessment of the long and eventful career of Professor Prakasa Rao at a time when the light is still glowing brightly and more so, when it is made by his student twenty years younger in age, who owes much to the volatile brilliance of his mentor. Nevertheless, Prakasa Rao's professional life, as a leader, has unfortunately come to an end and there is some merit in looking back at his contributions from the vantage point of personal and temporal proximity. A more distant and perhaps more objective assessment may be made by future generations of Indian geographers.

A scholar is a product of the times as much as of the geographical milieu. Both these aspects are important to the assessment of Professor Rao's contributions to Indian geography. His career forms a bridge between what may be termed as traditional and modern geography; a stage of transition that has materialised during the post-independence period. Further, Rao's career has to be judged in the light of the emergence of Indian geography, as different from colonial geography, which involved a new generation of Indian geographers whose mantle of identity underwent significant change. To be modern and to be Indian were the twin focii around which Indian geographers had to function in the post-independence period. Professor Rao has shown the way to successfully orbit around these focii.

Prakasa Rao is a pioneer in the application of advanced statistical methods and he has successfully demonstrated how these may be employed in an innovative manner to gain new insights into aspects of regional structure and development, problems of urbanization and intra-city structure, analysis of macro-spatial patterns and the spatial dimensions of socio-economic change. The choice of tools and choice of problems of study have an overriding importance from an Indian point of view. Prakasa Rao's perspectives for Indian geographers stand in sharp contrast to western prescriptions for the ex-colonials. A recent American formula would like Indian geographers to spin yarns around the elements of caste, culture and history as central themes in Indian geography. Hopefully, Indian geographers will heed Prakasa Rao's well articulated priorities for geographical research in India.

Where does Indian geography trace its roots ? Perhaps, we should go back to the 1920s and trace the contributions of Subramaniam in Madras, Sohan Lal in Lahore, Chibber in Varanasi, and Tabir Rizvi in Aligarh. The first two did not receive any education abroad and their efforts bore an indigenous stamp. The first professional journal of geography was the work of a school teacher. The articles written in this journal in the twenties and thirties show a robust concern for the understanding of patterns of agriculture and its relation to environment, aspects of regional economics, ethnic and social diversity and the evolution of places of importance. Description rather than analysis or synthesis dominated the approach of the self-made geographers of this time. The next phase, however, destroyed this rather idyllic setting. It was the period of emergence of western educated Indian geographers at the helm of geographic education and research. This period, covering the forties and early fifties, reduced Indian geography to a mere intellectual shadow of the west. Regional studies and the study of urban places followed well established paths prescribed by the French and British geographers. Independence and the growth of universities and other institutions of higher learning and research changed the content, but not necessarily the approach to geographical studies in India. It was at this stage that Prakasa Rao and his peers in the field tried to cope with the issues of regional and national development. It was not easy to find the right approach to these problems. Inevitably western, particularly European, approaches were scanned and new combinations of techniques of regional survey and analysis were evolved. Prakasa Rao's numerous published articles of this period show the increasing sophistication of approaches and tools of analysis—from simple statistical methods to more sophisticated models and computer applications. Regional development and urban studies became the focal points of geographical research in India. The number of urban geographers and regional development specialists in India today outnumber all others by a margin of 2:1. The earlier concern with agriculture and landuse and regional studies, though still important, has been overwhelmed by the new focii of geographical research. Prakasa Rao had a key role in bringing about this transformation in Indian geography.

Here, perhaps, a brief resume' of Prakasa Rao's professional career would be pertinent. Born in the present city of Visakhapatnam in Andhra Pradesh, Prakasa Rao had his geographical education essentially in Calcutta. He was the first candidate in India to receive the doctoral degree in geography. His guide was Professor Chatterjee, the leading geographer in India at that time. Prakasa Rao's first professional appointment was at the University of Madras in the early 1950s, where he was first a Senior Lecturer and then a Reader in geography. The Department of Geography at Madras at that time was headed by the late Professor George Kuriyan, a leading personality in Indian geography. At Madras, Prakasa Rao had his initiation in teaching and research guidance at the post-graduate level. In addition, he became the editor of the *Indian Geographical Journal.* A major turning point in Prakasa Rao's career was his association with Professor Learmonth in the late 1950s which finally resulted in his moving to Calcutta as the head of the Regional Survey Unit of the Indian Statistical Institute. Though Prakasa Rao was primarily a research worker rather than a teacher, the Indian Statistical Institute did not prove to be an ideal venue for the flowering of his original ideas on regional development planning; he left the Indian Statistical Institute for a teaching career in Osmania University, Hyderabad, in the early sixties. The home soil proved to be more fertile for the development of Prakasa Rao's penchant for research. For the first time, he had the opportunity to head a University Department of Geography and he did impart a new sense of direction and vitality to the Department. Prakasa Rao also has the unique distinction of being India's first professor of geography who had had his entire training within the country itself. From the interaction with Learmonth and the association with the pioneering work on regional surveys for planning in Mysore, Prakasa Rao emerged with a strong interest in urban and regional development and planning.

From Hyderabad to Delhi in 1966 was basically a professional upliftment from the regional to the national level. Delhi University, then at the climax of its academic excellence, provided a better atmosphere for research. Here he also met and developed associations with sociologists, notably, Professor M.N. Srinivas. His most important work during this period was a project on district level planning and the Vijayawada urban survey. In addition he wrote several important papers on settlement systems and regional disparities in India. All these were carried out within a span of six years. There was a substantive effort, at this time, to push the application of advanced statistical methods as an essential tool of his research. Further, electronic computers were used for the first time in geographical research. This was made possible by the fact that Delhi University was among the very few institutions of higher learning with computer facilities. His forays into quantitative analysis were largely supported by M.N. Pal, R. Ramachandran and V.K. Tewari who were all involved in the various research projects that he directed at this time.

In 1972, Prakasa Rao moved again from Delhi University to the Institute of Social and Economic Change at Bangalore. Perhaps his best research effort was accomplished during this period ; for, at this stage Prakasa Rao had all the time to concentrate on research. The Bangalore Urban Survey represents a major advancement in inter-disciplinary research in India. This work was followed by a study of urbanization in India which summarises his work on urban problems.

Prakasa Rao's career epitomises the story of Indian geography during the three decades after independence. A person of restless temperament, Prakasa Rao moved from place to place and from one institution to another; thus, weaving a pattern, both spatial and non-spatial, of his own. No other geographer in India has shown the same degree of spatial and professional mobility. Patterns and mobility were in fact always key words in Professor Prakasa Rao's academic vocabulary.

A salient feature of Prakasa Rao's contributions to geographical research is his emphasis on techniques and research methods. In this respect, his contributions to Indian and world geography began in the early fifties with the use of factor analysis in the study of the regional structure in Madras state. This was made possible by his close association, during his student days, with Professor C.R. Rao, an internationally renowned statistician. Since then, Prakasa Rao has maintained a strong predilection for technique-oriented research. He never shied from learning about the actual application of complex statistical methods. In fact, he demonstrated the different ways in which techniques could be productively used to gain new insights into geographical problems. Prakasa Rao was not only the first geographer in the world to use factor analysis but even more significantly, he used this technique in a wider range of situations—from the investigation on regional disparities, measurement of the importance of central places, study of regional and urban structure. It is this aspect that highlights the originality of his approach and it will be some time before his penchant for the use of techniques in research methodology is fully appreciated and emulated by the newer generation of geographers in India. After all his factor analytical study of 1954 did not attract attention until it was used with rather stunning effect by Brian Berry in the sixties !

Prakasa Rao had an equal, if not greater, interest in cartographic techniques. His fondness for map designing exceeded his love of statistical methods. Numerous examples of meticulously designed maps can be found in his published work; but even more importantly, he used maps as essential tools of research. In every research project, he devoted a lot of time to map analysis. In particular, the village level mapping schemes adopted first in the regional inventory of Muzaffarnagar District and later in the study of the Bangalore Metropolitan Region merit mention. He has a rare ability to combine within the same map several related spatial dimensions and highlight the inter-relationships of phenomena in space. He achieved this by careful choice of shading schemes and colours, with inset maps and diagrams carefully placed within the map area to provide balance and aesthetic form. It was always a pleasure to hear him lecture on the basis of map slides, which he used with great sensitivity and thus generated rare insights into spatial patterns and inter-relations.

As a teacher, research guide and colleague, Prakasa Rao set very high standards of devotion and diligence for himself and his students. Work was almost an obsessive passion with him; this he carried out with rare determination, even to the detriment of his personal and family life. The burden of his tenacity fell heavily on his enormously patient wife. Mrs. Rao, a true Indian wife steeped in South Indian tradition, was extremely warm and affectionate to her husband's colleagues and students; for Rao's devotion to work was such that his house also served as a second work place, which his close associates and students often visited. It is not an exaggeration to say that Prakasa Rao was an untiring student and researcher who devoted all his waking hours to work. In the middle fifties, when I was his student at Madras, he gave the impression of a task master; his lectures which bore the distinct stamp of very hard preparation were not easy for the light-hearted student and he expected all his students to read voluminously. Prakasa Rao was never casual in whatever he did; he took everything seriously and he expected every one around him to do the same—a proposition that would most clearly be unpalatable to the Indian geography students of the eighties.

Professor Prakasa Rao's major contributions to Indian geography are in the area of urban and regional development and planning. Interest in both these areas grew in the fifties when he was still working at the University of Madras. Both were further enhanced by his association with Learmonth. Prakasa Rao's first major study was on the towns of Mysore State (now Karnataka) in which he investigated the functional importance of towns through primary data collection and analysis. This study, in fact, was done long before the Indian geographers including Prakasa Rao became aware of Christaller's central place theory. This study is still a very valuable contribution not only in terms of the methods of data analysis and presentation but also in terms of locational analysis of inter-urban patterns at the meso regional level.

Following his work on towns of Mysore State, Prakasa Rao prepared the ground work for his later studies on regional development. He wrote a monograph on regional planning in the early sixties highlighting the contributions of European regional planners and their relevance to India. Apart from probing into certain aspects of regionalisation for planning and the study of regional differences in India, Prakasa Rao could get an opportunity to work on a major project on regional survey for planning only after his arrival in Delhi University. Here, in his first major research project, involving a team of geographers and statisticians, he examined in depth the regional structure of Muzaffarnagar District. This work has since been published and is entitled *Development Strategy for an Agricultural Region.* To all those who are concerned with the geographical component in district level planning, this study ought to be an eye opener and a model. The study has all the ingredients of a classic—depth in terms of village level analysis and profuse use of primary data ; range in terms of selective coverage of physical attributes, agricultural infrastructure, population and settlement structure ; variety in methodology, in terms of cartographic and quantitative analysis; all this topped by a chapter that attempts a synthesis of structural components and examines their relevance to planning.

Regional development, even in a rural context, inevitably leads to the study of the role of urban centres and finally to urban planning. A chance encounter with Dr. K.L. Rao, then a senior cabinet minister, led to Prakasa Rao's involvement in a major urban study ; this time, the venue of research was far away from Delhi, but nevertheless within his native state of Andhra Pradesh. The socio-economic survey of Vijayawada is a novel research project in at least three major respects. First, it involved actively geographers, sociologists, economists and town planners in a collaborative research project. This in itself represents a milestone in geographical research, since for the first time, a multi-disciplinary team was being led by a geographer. This experience had both its pitfalls and strengths. The second aspect has to do with the rigorous use of spatial sampling methods and the use of structured interview schedules in urban geographical surveys. A sample of 1,075 households were interviewed to generate information on a variety of structural and behavioural aspects of urban life. Thirdly, this study used computer data processing and analysis for the first time. In fact, the computer was used not only to produce tables, but a wide variety of maps including flow maps, proximal maps, and choropleth maps. This study forms the foundation for the more elaborate study of Bangalore.

Prakasa Rao moved away from Delhi University at a time when the university had started developing cold (democratic) feet and he was sensitive enough to recognise the changing times. At Bangalore, he developed a fruitful partnership with M.N. Srinivas, the eminent sociologist, and this led to the conception and execution of the Bangalore Urban Survey Project. The report of this study has since been published. This study is notable for the depth and variety of its analysis. It can be said with confidence that the Bangalore Urban Survey of the 1970s is by far the best urban study so far, in relation to the 30-odd major urban surveys that have been carried out by social scientists in India during the past three decades. The main advantage of the Bangalore survey was that the research team was working full time from within the city itself ; this, unfortunately was not true of the Vijayawada project, which was rather executed by remote control from Delhi and Hyderabad. This clearly accounts for the difference in quality between the two studies.

In the penultimate stage of his career, I am inclined to think that Prakasa Rao's interest and perhaps even his faith in regional planning waned rapidly. He saw the general importance of regional planners (for that matter all planners). Increasingly, he became an urban geographer *per se*. His recent work on *Urbanization in India : Spatial Dimensions*, reflects the change in his thinking and his priorities, for research. This work is in many ways a sort of summing up of earlier forays in urban research; but it also indicates a distinct move from micro or meso level to macro level generalisations. It is a pity that although in his life, Prakasa Rao had to contend with the arduous task of micro and meso level investigations, there was not enough time for a macro or all-India level study of urbanization to bring out the relevance of the insights gained at the micro level to the issues of national importance.

Two issues not totally unrelated to each other are important to the development of Indian geography. The first has to do with the relative role and inter-relationship of micro level studies and macro level generalisations. A measured jump from the micro to the macro level is a necessary stage for the evolution of an indigenous geographical base. Prakasa Rao has shown the way to handle the micro level analysis and he has explored macro level theorisation. Further work in this direction is necessary. The second issue relates to the western influence on Indian geography. Although European and American degrees are no longer convenient passports to professional advancement, there is still a considerable hang-over of intellectual dependence. The judicious borrowing of macro theory from the west and its application to India will not be adequate for the future; nor would the impudent induction of ideology help in solving problems. It is my opinion that Prakasa Rao's unique standing in Indian geography is based on an implied need for keeping out of doctrinaire ideological traps—western or even Indian.

Prakasa Rao has always been an independent thinker. He is more research oriented than most, if not all, of his peers in the field. His 'Andhra' temperament, which few people understood properly, matched by his Bhagirath-like devotion to Indian geography, has rightly earned him a high standing and reputation. His published work is indeed meritorious and will continue to inspire Indian geographers for decades to come.

4

CONTRIBUTIONS OF PROFESSOR V.L.S. PRAKASA RAO IN THE FIELD OF GEOGRAPHY IN INDIA

SUDESH NANGIA

PROFESSOR V.L.S. Prakasa Rao has been the precursor of studies in regional planning in the larger ambit of geography in India. He was one of the few to realise the importance of an inter-disciplinary approach in geography and regional planning and hence his team of researchers included statisticians, economists, econometricians, sociologists, town planners and demographers apart from geographers. His works have been deep-rooted in interdisciplinary and applied aspects of geography. His contributions to geography are reflected in his studies from *Towns of Mysore State* to *Indian Cities* and in the research work of his own students, fellow colleagues and their students.

Basically interested in regional planning, Prakasa Rao has constantly been thinking of envisaging a scheme of planning units for the country. Though he did recognize the importance of administrative units for the purposes of data and plan implementation, he could never reconcile himself to their being taken as planning units for the formulation of regional development strategies. His planning units were to be geographical—having resource homogeneity. But in his subsequent studies, he explained that homogeneity, centrality and nodality constitute the mutually inclusive concepts for regionalisation for planning.

His ideas in part got shaped in the Census of India

Monograph Series 1961 when *Economic Regionalization of India : Problems and Approaches* was released. P. Sen Gupta's article on the "Formation of Economic Regions by Population Characteristics and Resource Development" in this volume apart from others, was a partial achievement towards the identification of planning units. Further, K. V. Sundaram's work on the "South-East Resource Region" at the Town & Country Planning Organisation during 1966-68 was the culmination of his ideas on the identification of resource regions for planning. In addition, through his own works from *Planning Regions in Mysore State* in 1960 to *Urbanization in India : Spatial Dimensions* in 1983, Rao's ideas on regional planning and planning regions have grown from infancy to primehood.

Though the study of small and medium towns in urban geography by the U.N. agencies became a focus of attention only recently, it was mooted by Prakasa Rao as early as in 1963 when he undertook a project on the "Small and Medium Towns of Telangana" sponsored by the Planning Commission. He was one of the first to highlight the significance of such urban centres for a balanced regional development.

Prakasa Rao has been a pioneer in developing the concept of the urban-fringe in the Indian metropolis. From Sudesh Nangia's work on *Pattern of Rural Settlements in Delhi Region—A Geographical Analysis* (1972), where the idea of the urban-fringe/rural urban fringe was mooted, he worked meticulously on it till it received its final shape in Prakasa Rao and V.K. Tewari, *The Structure of an Indian Metropolis—A Case Study of Bangalore* (1979). Along with fringe, his interests were also focused on the rural settlements within the metropolitan range of influence which faced the onslaught of city life and culture and went through a process of structural, economic and social transformation. These settlements, popularly known as 'urban villages', became the special focus of study through the works of K.V. Sundaram, V.K. Tyagi on village Kotla Mubarakpur (1971), and *Urban Village of Delhi: Structure, Pattern and Transformation* (1976) ; Bina Srivastva on *Rural-Urban Fringe of Delhi : Structural and Functional Patterns* (1976) and Sudesh Nangia on *Rural Settlements in Transitional Environment : A Case Study of Ber Sarai in*

Metropolitan Delhi (1981). All these urban villages incidentally formed part of the Delhi Metropolitan Region.

Along with the analysis of the structural and functional character of fringe-villages permeated the ideas on rural-urban interactions, intra-urban interactions, commuting and movement patterns which have been observed through several research studies supervised, encouraged and/or directed by Prakasa Rao. Sudesh Nangia's work on *Pattern of Rural Settlements in Delhi Region : A Geographical Analysis* (1967-1972), Rangamani Parthasarathy's *Commercial Retail Shopping Structure of Urban Delhi* (1967-1972), C.S. Yadav's *Spatial Pattern of Residential Landuse Structure of Urban Delhi* (1971-1976) and Prakasa Rao and V.K. Tewari's *The Structure of an Indian Metropolis, A Study of Bangalore* (1979) include important components of such interactions which point towards a functional interconnectivity of spaces within the conceptual framework of the spatial continuum.

The development of this line of research culminated in the exchange seminars on 'Rural-Urban Interactions' between Indian and British geographers. Prakasa Rao has been a great exponent of the use of statistical and mathematical methods of geography. Needless to say, when there was a wave of quantification in geography in the western world, Prakasa Rao was practising it within the country. He had already joined hands with mathematicians like Professor C.R. Rao and econometricians like Professor K.L. Nagar and was exploring the possibilities of a meaningful dialogue with them. In the process, several techniques like factor-analysis (used by Professor M.N. Pal, one of his Ph. D. students and now a Professor at the ISI, Calcutta) found their use in the studies in geography.

Prakasa Rao is one of the few geographers who emphasised the importance of thematic cartography as a substantive and technical means of research in geography. He always stressed and demonstrated the use of maps not only as analytical tools but also as tools to generate hypotheses, identify spatial processes and patterns, and identify development problems.

In the process, he trained several geographers to be good

cartographers and several cartographers to be good geographers. His dynamism is also reflected in the introduction of computer-mapping in Indian studies in geography which he initiated at the Department of Human Geography in Delhi University in 1968-69 for the first time. Through a team of research scholars and a colleague Professor R Ramachandran (also a former student) who had expertise in quantitative techniques and computer programming, Professor Rao managed to get a computer programme in mapping developed and a series of maps were prepared through this process facilitating research in the department.

One of the rare qualities of Professor Prakasa Rao has been his meticulous use of secondary and primary data. Having served in the Indian Statistical Institute, he visualised the problems and limitations of data collection by any individual or organisation. Hence, it was difficult for any scholar working with him to affix the blame on data in case of erroneous results. While in discussion with him he would immediately ask, "Given the data-constraints, what are your observations;" He encouraged supplementing the observations from the secondary data with the observations from the primary data collected from the field. Hence, the research studies associated with him or conducted under his supervision have a pattern of macro-analysis based on secondary data and micro-analysis based on field-data, each data set acting as a counter check on the other. His dictum is that the power of the data and the power of the techniques must be judiciously balanced.

His philosophy and the approach towards the discipline of geography is well illustrated in the following Ph.D. dissertations submitted under his direct supervision and guidance.

List theses with Scholar's name, Year of Submission and topic

Gangawar, R.S. (1971), *Urban Geography of Pilibhit*.

Hanspal, Ranjana (1978), *Urban Structure of Bangalore City—A Spatial Analysis*.

Mazumdar, Madhusree (1981), *The Structure of an Industrial City: A Case Study of Durgapur*.

Nangia, Sudesh (1972), *Pattern of Rural Settlements in Delhi Region—A Geographical Analysis.*

Pal, M.N. (1968), *Regional Analysis for National Development Techniques and Case Studies.*

Parthasarathy, Rangamani (1972), *Commercial Retail Shopping Structure of Urban Delhi.*

Patni, R.L. (1969), *Concentration and Dispersion of Major Industries in India.*

Tewari, V.K. (1973), *Statistical Analysis of Statewise Differences in Consumer Expenditure Pattern and Level of Industrialisation in India.*

Venugopal, B. (1969), *Urbanisation in Telengana.*

5

PRAKASA RAO, THE PLANNER

K. V. SUNDARAM

I CAN hear Professor Rao objecting to my title. He says: "I am a geographer, not a planner. I do not like getting into somebody else's shoes". Incidentally "getting into other's shoes" is a fond expression with Professor Rao who used it often to discourage those members of his profession who with a mere nodding acquaintance with other disciplines and a naive empiricism, attempted to parade as inter-disciplinarians. Yet at the same time, Prakasa Rao is one of the few geographers who strongly advocated inter-disciplinary thinking and studies and wanted geographers to combine the penetrating insight in their own specialisation in geography with an adequate grasp of the wider complex of knowledge afforded by other disciplines.

While straying into inter-disciplinary areas with his own "weaponry and tactics", Dr. Rao did face many complicated situations and occasional frustrations. But he persisted in his efforts and it must be said to his credit that he could bring about some cross fertilisation of ideas and techniques which are amply reflected in his writings. Indeed, in my opinion, these constitute his unique contributions to geography today. It is my earnest hope that someone someday will write an objective essay on his encounter with statisticians and economists at the Indian Statistical Institute. Andrew Learmonth's paper in this volume is a partial attempt to do just this.

While considering the contributions of Professor Rao to the field of planning, one must recognise that there are two broad groups of planners in the country. One consists of socio-economic planners who are a heterogeneous group, mostly ill-organised or at best organised on individual discipline lines. The other is a more organised or institutionalised group of physical planners. It is to his credit that Dr. Rao could interact with both these groups and was welcome in both the camps. Establishing a rapport and dialogue with the latter group of physical planners and winning their goodwill and respect is, in fact, very difficult. I spent the better part of my professional career in close association with this latter group and having gone through traumatic experiences am well aware of the difficulties in getting 'other viewpoints' accepted by this group. My essay on "Some Problems of Interdisciplinary Team Work and Research in Urban and Regional Planning" which is a chapter in my book on *Urban and Regional Planning in India* (Vikas, 1977) is a faithful record of my experiences while interacting with this group of planners. But Prakasa Rao did find acceptance among this group of planners, was highly respected by them and was a constant invitee to their annual town planning seminars. On every theme around which such seminars are organised, one invariably finds the geographer's point of view articulated by Prakasa Rao. Unfortunately, there is no full bibliographic account of these contributions made by him. Most of these papers have been published in *Town Planning*, the official journal of the Institute of Town Planners. My original plan was to prepare an annotated bibliography of all the research publications of Professor Rao for inclusion in this volume. My regret is genuine that I could not do so on account of constraints of time.

The advocacy of the regional concept and regionalisation for planning was one of Professor Rao's major preoccupations in research during the sixties. He generated a lot of new ideas on the subject and, what is more pertinent and significant in this context, he experimented with numerous regionalisation techniques himself. He was a pathfinder in this field and encouraged many geographers to do work in it. One important trait in his character as a researcher has been to stimulate and encourage fellow researchers who evinced interest in the same area of research. He did not consider them as his rivals, but rather as associates who complemented his own efforts. I recall the extraordinary interest that he took and the technical support and advice that he readily provided to me along with Professor Norton Ginsburg, when I attempted a scheme of inter-state regionalisation in the Town and Country Planning Organisation at the instance of C.S. Chandrasekhara, a doyen among town planners in India. In this attempt, some fresh ideas on the criteria of regionalisation emerged. Similarly, I can recall how Prof. Rao encouraged Dr. Reed, an American geographer, in a study of metropolitan interactions. Instances of this kind are numerous. The other assessments about Prakasa Rao included in this volume have also referred to this aspect of his personality. Prakasa Rao's original and inquiring mind has always been in search of new frontiers in geographical research. The result is that several indigenous research areas peculiar to the Indian context, which had not been researched upon before, have come to be identified. The urban village is one such area and Prof. Rao provided the initial stimulus to study this phenomenon in Delhi. My study of Kotla Mubarakpur in Delhi prepared for the first Indo-British seminar was greatly influenced by his thinking. Subsequently this paved the way for a full-fledged doctoral dissertation on the subject.

Prakasa Rao's thinking has always been well ahead of the times. I would particularly refer in this context to an off-beat exercise—in fact, a fall-out of the Mysore regional studies—his study on the reorganisation of district administrative boundaries. This was done during the 60s. It is only today that many state governments in India are seriously applying their minds to this important question, namely, the optimal size of a district for planning and administration.

While discussing the role of Professor Rao as a planner, one cannot fail to make a reference to the Muzaffarnagar study, which is remarkable for its in-depth analysis. Here he has demonstrated how with the insights obtained through micro analysis at village level, one can formulate a district plan on a realistic basis and influence macro (regional) economic policies. Unfortunately, this work has not received the attention it deserves among planners and administrators. The concept of decentralised district planning is now being looked at with great interest in the country and it is going to be one of the cornerstone strategies during the 7th Plan period. In this context, it is hoped that Professor Rao's contribution will receive due recognition.

In all that Prof. Rao wrote and advocated during the last three decades, his concern for objective analysis stands supreme. His latest book on *Urbanization in India: Spatial Dimensions* (Concept, 1983) starts with a quotation from Stopler, "Truth must come before originality." Nothing can sum up more aptly and more precisely Prakasa Rao's philosophy and attitude to Indian geography and planning. In fact, it is this search for truth and objectivity in research which characterises Prof. Rao as one of our unique geographers.

It is pertinent to make a few remarks in this context about Prof. Rao's style of writing. I would characterise his style as pithy, precise and pointed, with no verbiage or superfluity about it. Prof. R.P. Misra and I who collaborated with him in a joint work on *Regional Development Planning in India—A New Strategy*, (Vikas, 1974) found it tough going at every sitting with him. Prakasa Rao would not agree to the inclusion of any sentence or word which did not have empirical backing. Such was his insistence on analytical rigour that his research students always found it difficult to measure up to his expectations. But once they were through his mill, they acquired a certain stamp of scholarship and qualities of head and heart which stood them in good stead in their career. Many of his research students have expressed these sentiments to me in private conversation. I myself came under the influence of Prof. Rao as early as the 50s when I started my career as a college lecturer. Subsequently, when I moved into government service and was often entrusted with some challenging assignment or the other, I used to go to Prakasa Rao for advice and always looked up to him as my mentor and guide in several research activities. I have invariably been the beneficiary in this association.

Before I conclude this brief assessment, I should not fail to make a reference to two of his latest works—the Bangalore Study and the study on *Urbanization in India: Spatial Dimensions*—both of which represent the quintessence of his thinking as developed since the 60s. They contain many useful insights which are bound to influence the thinking on the applied aspects of Indian geography in the years to come.

Finally, I would like to add that this assessment of Prof. Rao as a planner is subjective and partial, not rudimentary and fragmentary, as any assessment of great minds by a single individual is bound to be. There is need for a fuller assessment by a group of geographers and other professionals in allied fields who have been influenced by Prof. Rao's thinking. This is a task which institutions like the ICSSR should initiate, because here is an inter-disciplinarian who has contributed much to cross disciplinary thinking among social scientists.

Prakasa Rao has yet some unfinished work to do. The experiences which he went through, the empirical studies which he made and the interaction and exposure that he had with the intellectuals of his age have given him a lot of ideas which remain to be churned in the crucible of his mind and turned into a consistent philosophy in geography and planning. During his extraordinarily busy academic life, there was no time for him to look back and reflect. Hence, Prakasa Rao, the philosopher is yet to emerge. His *magnum opus* is yet to come. Let us hope and pray that in the years to come, he will have the peace and the quiet, the health and the strength to mull through his own ideas and turn out a philosophical treatise for posterity.

SECTION TWO

6

INTER-REGIONAL DISPARITIES : Country Experiences, Policies and Strategies

K.V. SUNDARAM

THE spatial configuration of development in the developing countries is undergoing many changes under the impact of planned economic development.* In this context, the problem of inter-regional disparities is assuming great significance, calling for specific interventionist strategies and policies. The problem in the developing countries is vastly different from the one in the developed countries. Whereas in the developed countries, the problem is being tackled through some readjustments and minor re-orientations, in the developed countries major shifts in the development process are involved. These may include opening up of interior areas and initiating major resource and infrastructure developments along with shifts in the labour force, thus bringing about vital changes in the "geographical division of labour".*

Analyses of various country experiences serves to clarify the numerous policies and strategies involved in the planning for reduction of inter-regional disparities and to consolidate and redefine our thinking on the approach to the problem. The country experiences of five developing countries, Malaysia, Indonesia, Thailand, the Philippines and India, are considered in this paper. All these countries, with the

* K.V. Sundaram, *The Geography of Underdevelopment : The Spatial Dynamics of Underdevelopment*, Concept, New Delhi, 1983,

exception of Thailand, have emerged either in the 40s or 50s from a formal stage of colonial dependence, which has contributed, by and large, to their present spatial structure of underdevelopment. Generally speaking, since the sixties, they have been seeking to change their existing spatial structures. In their plan documents of the seventies, regional policies are becoming explicit in varying degrees. It must, however, be admitted that there are often glaring differences between policies, plans and practices in these countries. Till recently, regional issues were evaded by them because national (i.e. macro level) issues were dominant claiming their first attention and/or because the sub-national problems were of formidable magnitude and colossal complexity, placing severe strains on their slender personnel, institutional and financial resources. The nature of the problem, of course, differs from one country to another, although there may be parallels and similarities in certain respects among some countries.

THE PATTERN OF REGIONAL DISPARITIES

In Malaysia, the regional problem is not merely a spatial problem, but a racial, religious and cultural one too. In terms of levels of development there is, on the one extreme, the west coast of Peninsular Malaysia with urban growth and activity concentration in the Klang Valley and, on the other, an interior with vast virgin land and practically untouched resources. The poor or lagging regions are mainly along the east coast and partly in the North. Indonesia exhibits a *Plenum and Vacuum* pattern with developments concentrated in Java, Sumatra and Kalimantan and with the 'Outer Islands' remaining in a raw-material extraction and export stage of development. In Thailand, the problem is one of a Primate city and its hinterland characterised by backwash effects. In the Philippines also, development was historically concentrated in the island of Luzon, particularly Manila and its environs and a few urban centres in the other islands, leaving the rest of the country high and dry in the matter of development. So far as India is concerned, the spatial configuration of development shows that the under-developed areas are widely dispersed in the east, centre and middle-south of the country. One may identify this concentration as a belt spanning across the 'belly of India', with its extensions to the east and south. This is perhaps an over-simplification, for what is striking about the spatial disposition of backward areas in the country is their 'interspread' distribution, often in juxtaposition with relatively prosperous pockets. Thus the picture is one of pockets and zones of backwardness alternating with islands of prosperity together with their limited areas of spread effect of development. This is nothing like the north-south polarisation that one finds in Europe or the Americas. Perhaps the north-eastern part of the country is an exception, where an extensive area of under-development may be distinguished.

The spatial pattern of development in a country has a great bearing on the logistics of development, as it influences the policies and strategies that must be pursued to alter the course of development. In Indonesia and Malaysia, for instance, where virgin lands are available for settlement, population movement from one region to another by organising and injecting all pre-requisites of economic growth in the undeveloped areas becomes a relevant strategy for reducing regional imbalances in development. In India, on the other hand, where pockets and zones of backwardness are interspersed with areas of relative prosperity or spread effect, the problem is one of devising multi-pronged strategies for encircling or mopping up of these pockets.

In all these countries, inter-personal disparities are also striking and income distribution is highly skewed with the proportion of people living below the poverty line ranging between one-third and one-half or even more of their total population (Malaysia, Thailand, Philippines, India, Indonesia in an ascending order). Thus there are two aspects to the problem of regional backwardness and inequality in these countries : (a) Some regions are less developed than other regions either because their resources are underutilised or because they have resource constraints ; (b) there is also great inter-personal income inequality among people with the presence of a large proportion of the under-privileged population who have been marginalised during the normal development process. In order to tackle these twin problems, a target area as well as a target group approach are called for in the developing countries.

Regional development—strategies and policies

Regional development strategies in Indonesia are being pursued in three directions :

(a) By developing the undeveloped resources in remote regions through transmigration programmes ;

(b) through sectoral development strategies at the provincial level, involving construction of large scale projects such as dams, reservoirs, road networks etc ; and

(c) through a budget allocation strategy by means of subsidies and grants allotted to local governments.

Despite some limited achievements through these measures, the 'regional problem', still persists. The transmigration programme is evolving into a 'regional development' programme and in the settled areas, an integrated approach to rural development with 'Kecamatan' as the unit area of planning is being experimented.

In Malaysia, the national development policy is being guided by the objective of minimising inter-racial and inter-regional socio-economic development imbalances. The regional development strategy takes advantage of the availability of vast undeveloped areas. The primary strategy is either centred around land development (which is not simply land clearance and opening up of virgin lands, but also encompasses in its fold agricultural development as well as the setting up of agro-based and forest-based industries). The examples of such development may be seen in the Pahang Tenggara region or around some corner-stone projects e.g. irrigation by using them as a medium of integration for multi-dimensional activities (e.g. Muda Irrigation Project).

In the Philippines, the regional problem has been sought to be tackled by adopting a regionalisation approach. This is by dividing the country into different regions and by creating institutions and layers in government apparatus for the co-ordination of regional development effort. Thus the country has been divided into 13 regions and Regional Development Councils have been set up for each region. Yet, the Philippine experience shows that solving the 'regional problem' demands more than mere structures. It needs, besides decentralisation of powers, supportive budget, organisational efficiency and expertise in both planning and management.

In Thailand, the strategies for regional development include :

(i) A strategy for the intensification and diversification of the economy for improving the regional industrial mix and export base ;
(ii) a strategy for decentralised urbanisation ; and
(iii) a strategy for decentralised infrastructural and social services to the rural areas.

Because of the low base-level of development of the rural areas, simultaneous 'extension strategies' through local level planning and development are being attempted. But the impact of these measures is slow, as there are a number of obstacles to be overcome such as local power structure, dependency syndrome and bureaucratic obstructions to decentralisation of power.

In India's repertoire of policies and strategies aiming at balanced regional development, there are a number of indirect and direct measures which are of great interest. Five major lines of strategy may be distinguished :

(i) An inter-regional allocation policy for the distribution of central assistance funds to the state governments which is governed by a formula tilted in favour of backward areas ;
(ii) incentive policies designed to direct investments to the industrially backward districts ;
(iii) action planning based on area/regional development approach to tackle specific problems in identified problem areas e.g. tribal areas, hill areas, drought

prone areas, desert areas and problem regions like the north-eastern region ;

(iv) integrated approach to local-level planning focussed on the district and the block ; and

(v) a basic needs strategy oriented towards the provision of minimum needs, so that the disadvantaged areas and groups may achieve parity with others in terms of social consumption.

A special reference to the approach to backward area development will be useful. This approach is characterised by three important elements, namely, selectivity, specificity or focus and dynamism or flexibility. An all-at-once approach to the problem has been discarded ; indeed such an approach has been found not feasible. Experience with the development of the backward areas has revealed that the approach initially has to be one of making a deep thrust into the specific problem area(s) identified as major hindrance to development thereby contributing to the first breaking-down of what have been identified as the key obstacles to development in the particular area. Thereafter, the development strategy must add to this other elements thus gradually enlarging the scope of development, and contributing to greater and greater comprehensiveness in planning and development. This implies a progression from the particular to the universal or from part to the whole. But such progression will not be along a straight line, so that no simple rules of progression are possible. For each problem area, the appropriate methodology and sequence of approach must be devised.

The Indian experience shows that the closure of the regional gap will be effective only in a highly decentralised approach through micro-level planning in which it would be possible to go area by area and deal with its development in a disaggregated manner taking advantage of local participation, local perception, local insights and additional resource mobilisation at the local level. Attention to the target groups—the people below the poverty line—must be part and parcel of such planning; otherwise the weaker sections will tend to get by-passed. In large countries like India, planning has not only to be at the micro-level, but also at multi-level so that proper "nesting" of plans takes place and distortions do not develop.

The Indian experience also shows that the removal of backwardness is not necessarily achieved through lumpy investments alone ; very often, by making several half-way arrangements and adjustments and filling up gaps which may involve only small complementary investments, it would be possible to step up the development effort in a backward area to a substantial degree. Increasing the utilisation of existing capacities, already created in the area, is one method ; yet another would be through deliberate efforts to capture the various inter-linkages and externalities in development. Improving management for better performance, for example, by adopting an appropriate management approach, is still another technique. Such interventions involving limited investments are likely to lead to visible improvements. Again, certain non-economic measures, like inducing basic social and institutional changes in the area, are very important for the development of backward areas. In fact, providing the appropriate social and institutional basis for development is regarded by some as the most important step in the backward areas development strategy.

Thus from the experience available in India, one may say that the method of planning for a backward area would be: first to define the local area for planning and then to identify an appropriate package of development programmes for the area. This would involve defining and locating the areas where (a) concentrated investments could be provided ; (b) utilisation capacities could be maximised ; (c) certain gaps could be filled by capturing economic interlinkages and externalities ; and (d) certain basic social and institutional changes could be initiated.

Having defined these parameters, it would be necessary to resort to specific household planning of families below the poverty line—the target groups—so that concerted effort is made to reduce the interpersonal income disparities within the backward area. Thus the backward area development strategy appears to be one of adopting an area specific approach combined with a functional approach to reach specific problems and target groups.

THE CURRENT EMPHASIS IN REGIONAL DEVELOPMENT POLICIES

A significant aspect of regional development effort characterising the developing countries since the beginning of the 70s seems to be the adoption of a new philosophy of 'integrated approach' to regional/rural development. This integrated approach is a strategy that seeks greater and more effective co-ordination among rural development programmes and projects through a consolidated and synchronised development effort in the different sectors of the economy in a given geographical area to attain economic development and social equity. It is too hasty to conclude that this new approach has been fully realised in the actual policy implementation in these countries. The 'integration' implied in this approach has been found to be difficult of realisation by governments. The country experiences in this regard point to some 'prerequisites' that are called for. These are :

(i) the need to mobilise and train local people prior to or in parallel with the implementation of regional development policies ;

(ii) the need to pay some attention to the institutional dimensions of regional development. For example, the land development schemes in Malaysia, the Compact Farming Programme in the Philippines and Transmigration Projects in Indonesia have afforded clues in this direction ;

(iii) the need to bring about adaptations in development administration. The ineffective and inefficient administrative organisation, in particular at the local level, has been realised as the single most important obstacle encountered in implementing regional development policies.

SOME EMERGING CONCLUSIONS FROM MULTI-NATIONAL COMPARATIVE ANALYSIS

In India, as well as in the other developing countries, where regional disparities are in evidence for various reasons—historical, socio-economic and political—a number of indirect as well as direct policies are being pursued in an effort to keep down inter-regional disparities within certain "tolerance limits", as otherwise their persistence or acceleration may lead to serious repercussions such as national disintegration or political instability. The policy instruments devised by governments to reduce inter-regional disparities range from indirect to direct and from selective to comprehensive approaches. They include—

(i) capital and technology transfers ;
(ii) incentive policies for agricultural and industrial development ;
(iii) land development and resettlement with a package of incentives ;
(iv) regionalisation strategy ;
(v) integrated development focused on delimited small areas ;
(vi) target group approaches ;
(vii) bottom-up strategies and decentralised development ; and
(viii) comprehensive regional planning approach.

An aggressive policy aiming at a speedy closure of the regional gap would mean a highly accelerated development of backward areas in the country, which can have the effect of pulling down the national growth rates of the economy. In the case of developing countries, therefore, a temporal sequence of development must be envisaged. That is, the development of underdevelopment must be seen as different stages of transition growth.

This 'stages approach' to backward area development will have its ultimate goal in perfect inter-regional equalisation. Conceptually, the movement towards this ideal goal may be

Paradigm of shifts

Development aspects	Stage I	Stage II	Stage III
1	2	3	4
Economic	Primary production in settled areas Opening up of new lands and frontier development	Intensification and diversification of the economy Application of science and technology. Diffusion of innovations End of 'frontier development'	Towards capturing regional comparative advantage and interlinkages Interdependent development through interregional specialization
Spatial	Spatial structure characterised by features of colonialism	Changing spatial structures in varying degrees of 'dependency syndrome' (Colonial structures persisting through internal colonialism)	Towards spatial integration
Population	Redistribution of population—internal migration and adjustments	Internal migration of different orders	Population adjusted to 'Territorial Division of Labour'
	Low base of skills	Beginnings of skill development	Acquisition of advanced skills Intense mobility of labour

1	2	3	4
Social and political	Economic feudalism and power structure of the elites Excessive class, group of ethnic differentiation	Shifting power structure in community. Growth of people's institutions trends towards social mobilisation	Democratic Decentralisation
		Class, group or ethnic contributions Regional claims and conflicts	Integration of society
Administrative	Centralized administration Top-down structures	Executive decentralisation i.e. de-concentration	Decentralisation with delegation of powers and authority Towards integration of development administration with democratic institutions
Nature and extent of inequalities	C. Sharp, exhibiting features of unbalanced use of natural resources (eco-destruction)	G. Un-balanced development; the need for eco-development (discordance)	G. Towards maximization or regional potentials (accordance)
	D. Commencement of inequalities	D. Sharp and fluctuating	D. Towards a pattern of 'inequilibrium within tolerance limits'
Nature and extent of inequalities	SE-Sharp	SE-Increasing	SE-Declining; Towards an egalitarian society

1	2	3	4
Process :	Selective differentiation of problems and determining the areas for priority intervention based on region's comparative advantage	Expansion (widening of the bases of interaction) and consolidation. Iterative process of differentiation. followed by partial integration	Integration

Note : *G*—refers to Geographic inequalities; *D*—refers to Development inequalities; *SE*—refers to Socio-economic inequalities.

envisaged as a 'paradigm of shifts' in three stages as given below.

It must be understood that in the paradigm, events are presented in a more systematic fashion than they are likely to occur in the real world situation. The conceptualisation attempted here is based on the belief that development would be totally "controlled and manipulated", that it would be backed up by unflinching political will and that there will be uninterrupted continuity and consistency in government's policies. These are assumptions difficult to come by. The first and second stages in the paradigm are derived from the development experiences in the developing countries and, therefore, are reasonably realistic. The third stage, however, is an abstraction, a conceptual purity and an elusive goal. There are several imponderables in the situation which make the realisation of this stage of development difficult for any country. Other concomitant approaches that could complement the 'stages approach' towards the development of backward areas would be—

(i) Adopting a typological approach to the problems of backward areas. This would be by classifying these areas on problem lines and adopting area specific and problem specific strategies for development.

(ii) Adopting suitable policies for arresting 'leakages' from the backward area and thus changing an 'extrovert' economy into an 'introvert' one. This is an approach meant for 'internalising' the development process within the backward area and ensuring maximum gains to it.

(iii) Adopting a comparative advantage principle in planning for the backward area in order to promote 'inter-dependent' development, among regions. This will tend to bring about mutual complementary relationships between the backward areas and the relatively developed areas and thus avoid or reduce exploitational tendencies.

Several policy instruments were mentioned earlier in this

discussion which have varying potential under different circumstances to influence 'closing the gap' in development among the regions in the country. With an appropriate 'policy mix', it is possible that maximisation in 'closing the gap' could be achieved. However, what such 'policy-mix' should be in a given context is a particularly intriguing question. In fact, finding the right answer for it constitutes the real challenge for the planner today.

7

REGIONAL DEVELOPMENT : Some Lessons from the Sudan

H. R. J. DAVIES

"PLANNING should be seen as a means so to organise human society that it can adjust itself to the changing socio-technical environment and can use this environment to maximise the welfare of its members" (Misra, Sundaram and Prakasa Rao, 1974) ; and again they write : "Development planning is designed to generate structural changes in society in order to facilitate the growth of the national economy". Prakasa Rao and Bhat (1964) have also written : "The regional approach to resource development is essential in India where regional contrasts . . . . are quite sharp". India has been in the forefront of development planning among developing world countries.

In December 1933 the Legislative Council of the United Provinces passed a Resolution recommending the formulation of a Five-Year Plan. In 1937 a Report on Economic Planning in the United Provinces was presented which concluded that planning was better than "piecemeal and isolated developmental activities" (Hanson, 1966). From the inauguration of the First Five-Year Plan in 1951 the need for democratic planning by consensus and for engaging even the lowest levels of administration within the process to ensure an unanimity of purpose was emphasised. The need for social as well as economic development too was stressed and, because of the size of the country with its environmental and cultural complexity, planning from the start was to be a combination of centrally and state organised initiatives.

In practice due to the inherent contradictions in these stated aims and pressure of circumstances one aim had to be subordinate to another. Planning tended to be economic rather than social : the economic aims were now to lead to 'social justice'. Planning also tended to look at national aim and to be sectoral rather than spatial, so that regional planning, except for special target areas such as Delhi and Calcutta, was largely ignored. The democratic consensus was liable to be converted into uniformity dictated from the centre to the region.

During the 1970s 'multi-level' planning began to be emphasised and it became no longer quite so true that "The state plans were nothing but catalogues of demands. Grass-roots planning, which was supposed to form the basis for national planning, has been conspicuously absent" (Misra, Sundaram and Prakasa Rao, 1974). At the same time some of the lack of realism that resulted from a failure to understand both group and individual motivation has begun to disappear with an increasing awareness of the importance of regional cultures (Sundaram, 1977). "When will economists and the government realise that economic reforms need to take into account the social constitutions within which economic factors operate ? Only in India could the government launch on a Five-Year Plan without consulting a single sociologist who has a first hand acquaintance with the social institutions of the people" (letter in the *Economic Weekly* for 4 October 1952 quoted by Hanson, 1966, p. 102). It is unfair to say "only in India", but undoubtedly the point had and still has some validity.

To plan from the centre, to tell the people what is good for them, to believe that as all men have a theoretical equality then they will all react in the same way is undoubtedly a comforting set of guidelines for a central government wishing to move quickly forward in development. It implies that there are short cuts to avoid a great deal of painstaking research and investigation and suggests that blue-prints for particular programmes of change exist. Many countries such as Tanzania and Algeria have consciously chosen to walk down such a road and have run into great economic and social problems, more particularly in the rural sector. In the industrial sector where all is new and the programme starts from scratch blue-prints may be useful, but by contrast in the rural sector the planner has to start with an existing situation within a highly developed socio-cultural system which is at the heart of a people's identity.

Like India, Sudan is a very large country with considerable environmental and cultural complexity. The far north is one of the most arid regions on earth, whilst in the far south equatorial forests can be found. Culturally, the more 'advanced' Moslem north contrasts with a less "advanced" and somewhat Christian south, whilst its total area (2½ mn sq. km) is equivalent to more than three quarters that of India. One very significant contrast with India is its overall population density of 8 persons per sq. km compared with India's 210. Development plans for India have been handicapped by the high rural population densities coupled with a rapid rise in rural and urban population, whereas in the Sudan, in spite of a current annual increase in population of about 2.6 per cent, her rural development programmes have suffered from a shortage of manpower with long lines of communication generating relatively little traffic to pay for their up-grading and maintenance. There is hardly a tarred road in the country outside town apart from the Khartoum to Port Sudan road and a very limited network in Gezira and White Nile Provinces. The railway remains unballasted single track with a speed restriction never exceeding 40 km p.h.

The history of planning in the Sudan also has many parallels with India. Immediately after the Second World War the Sudan Government set up a Development Priorities Committee which produced the 1946-51 Development Programme, an *ad hoc* collection of 200 projects. Similarly, the 1951-56 Development Programme was little more than a collection of projects. A policy was first proposed in the "Ten-Year Plan of Economic and Social Development, 1961-62 to 1970-71". In spite of its title it was essentially an economic plan drawn up by the Economic Planning Secretariat in the Ministry of Finance and Economics (Abdel Wahab, 1976). The main social concerns were to improve health and educational services by providing funds from the centre for these purposes. In spite of the size and diversity of the country very little attempt was made to take into consideration the characteristic cultural differences between one region and another. The only element of regional planning lay in decisions to site projects in particular areas, decisions based upon purely economic and political criteria.

Due to political instability, this Ten-Year Development Plan had little real chance of achieving its aims. It was replaced by a Five-Year Plan (1970-75) which was followed by the current Six Year-Plan (1977-78 to 1982-83). All these plans have emanated from the centre with the various regions presenting requests for various development projects with little or no attempt at a coherent policy for each region. "Many of the projects were conceived at the national level, which in many cases failed to fit local conditions" (Arifi, 1978). This is surprising, because from 1969 there was a distinct change in government policy with considerable efforts to involve people with development through the Sudan Socialist Union. This process was strengthened with the People's Local Government Act (1971) with one of its main objectives "to create channels for mass participation in decision-making, and to give local councils the right to prepare their own projects and plans and to seek local or other sources of finance for implementation" (Arifi, 1978). Over 5,000 local councils were set up and grouped together into 18 provinces under a four-tier hierarchical system. True grass-roots participation has failed to materialise. Almost everywhere in the country the regional plans were handed down from above, from central government to Province and so on down to village council level. Nevertheless, the importance of regional planning was presented in the 1971 Act and was reinforced at a national level with the Addis Ababa Accord (1972) which set up a Southern Autonomous Region composed of the Provinces of Bahr-el-Ghazal, Equatoria and Upper Nile.

During the 1980s the importance of the region has been strengthened with the creation of five regional assemblies in the north, and in May 1983 it was decided to extend this principle to the south with the creation of a further three regional assemblies. The nine new regions will be responsible for planning and its funding in their own areas within parameters laid down by the central government. Two problems with the new system centre on the excessive bureaucracy generated by having a six-tier administrative system from the central government down to the village council and the difficulty of finding sufficient numbers of competent civil servants to man the regional ministries as well as the national ones.

As in India, so in Sudan, planning has been essentially a central government enterprise with planning handed down from above ; it has been economic planning with the social element covered by education and health programmes ; and the grass-roots participation has remained theoretical rather than practical. It has also been sectoral and project orientated rather than regional in its concepts. Recent changes in policy stem at least in part from a re-assessment of the country's achievements in development in the light of the present economic crisis.

Over the last 30 years Sudan has placed most of its development efforts into the improvement of agriculture. Low population density, wide areas of underused lands, lack of fossil fuels and a reputation for high quality cotton production encouraged this policy. Expanded agriculture would boost exports, reduce imports by import substitution and provide for the development of cotton, oil seed, leather and similar agriculturally based industries. As a result, the areas under irrigated and rainfed agriculture have both increased five-fold though the population has only increased by 75 per cent during the same period.

Yet all is not well with agriculture. Government estimates suggest that yields per hectare for sorghum, the country's staple food, fell by 14 per cent during the 1970s. In 1981, Sudan harvested its smallest cotton crop for 40 years (and this crop normally accounts for 60 per cent of the country's export earnings). In the same year the Mechanised Farming Corporation, which operated the government's mechanised farming schemes, was wound up in face of unsupportable financial losses. In 1979 a survey of eight major agricultural development schemes concluded : "No project can be called economically sound from the private and public point of view" (Thimm, 1979). It also concluded that five of the eight had serious shortcomings in terms of "social acceptance" and that none were without social shortcomings. In 1982 Sudan's external debt amounted to over U.S. $ 5,200 million and the balance of payments deficit for the year was U.S. $1,000 million, so that a re-scheduling of Sudan's debt repayments was precipitated.

The current world economic crisis, rising oil prices, administrative and political shortcomings, weaknesses in infrastructure, lack of realism in planning and a chronic lack of foreign exchange for essential purchases are all important aspects of the problem, but even if all of these could be overcome the problems would still not be solved. This paper is a plea for regional planning, within a national framework, that recognises local socio-cultural differences, and it follows up the conclusion : "Wherever possible plans must be prepared with the people not for them" (Thimm, 1979).

Among the agricultural development schemes launched in recent years in the Sudan three will be examined, viz., the Khashm el Girba irrigation scheme (200,000 hectares) on the river Atbara, the Gummuiya pump scheme on the White Nile (3,500 hectares), and the Habila and associated mechanised agriculture projects in the Nuba Mountains area of Southern Kordofan (over 200,000 hectares.) (Map. 7.1).

Khashm el Girba was developed between 1964 and 1970 and was intended as a resettlement scheme for Sudanese Nubians whose homeland was to be flooded by the filling up of the reservoir behind the High Dam in Egypt; and to resettle the local nomadic groups integrating them into the national economy. At the same time it was envisaged as a scheme to improve Sudan's balance of payments. Cotton was to be produced for export and wheat and sugarcane were to be cultivated in an effort to reduce the annual import bill for these commodities. At the same time groundnuts were also chosen to help maintain soil fertility, for local processing for vegetable oil and for the export of oil and cake.

Regional Boundary

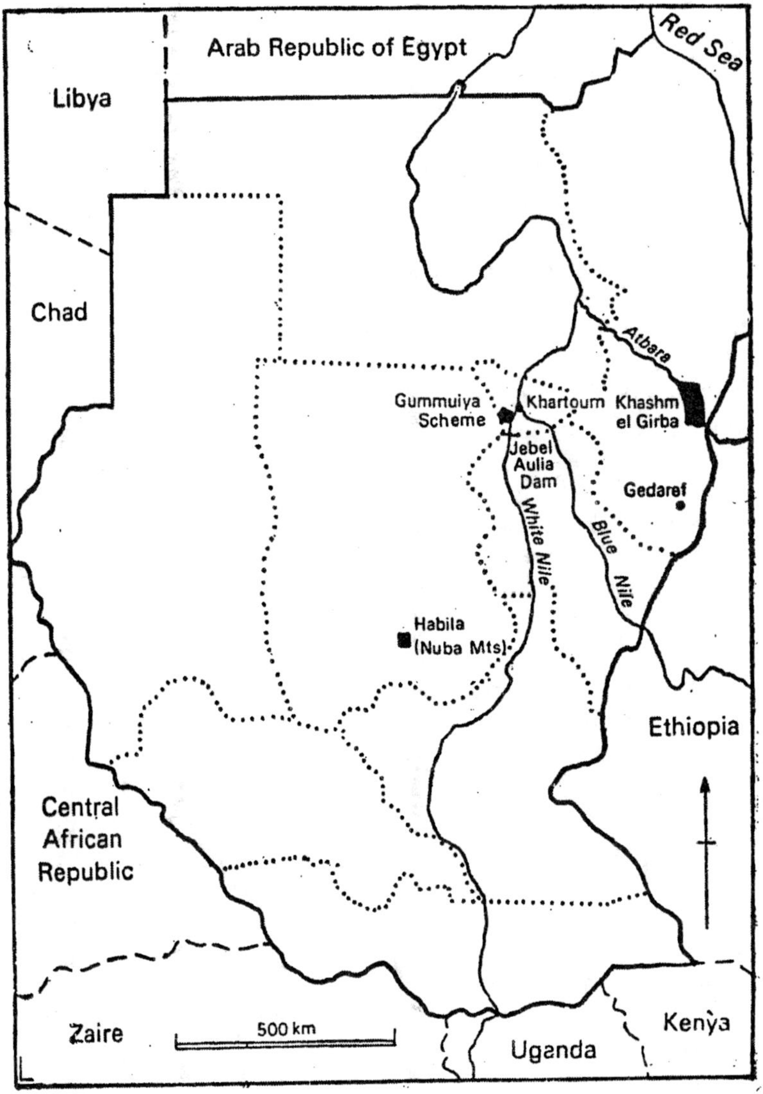

Fig. 7.1 : Sudan

The general organisation of tenancies, cultivation practice and profit sharing between the various partners was copied from the successful Gezira Scheme. As in the Gezira, willingness to grow cotton was a pre-condition for having a tenancy. Government accounting has recently suggested that the Khashm el Girba Scheme is being subsidised by the rest of the economy even on current account. Furthermore, the scheme has failed to meet the expectations of the Nubians and has proved unattractive to the local nomads. The tenants complain about the management and the management complains about careless, uncooperative, lazy and absentee tenants.

One basic problem has been the transfer of a 'blue-print' from the Gezira area. Not only was this scheme originally conceived in 1925 and has received little basic modification since, but it was also developed among people with strong agricultural traditions in an area needing changes in traditional practice to cope with a rapidly rising population on arid marginal lands. Though the introduction of irrigation meant a revolutionary change in life style, the people of the Gezira had already begun to perceive that changes were necessary.

In the Khashm el Girba region the position was (and is) different. Here, the local people were essentially nomadic with their way of life centred on camels, sheep and cattle rather than on their crops and there was a little perceived need for change among them. Furthermore, the relevance of the Gezira blue-print developed by a colonial administration was being brought increasingly into question during the 1970's and the Gezira itself was the subject of a national inquiry during 1980 and 1981. Again, when cotton was introduced into the Gezira there was a good demand for long staple from the colonial power and international prices were high. This was no longer the case when Khashm el Girba was developed.

So far as the nomads were concerned low returns from cotton were compounded by the decision to grow wheat, an unfamiliar crop and not their food staple, and to exclude livestock from the scheme. They are also aggrieved to see their ancestral lands being taken over by the government for the resettlement of Nubian strangers. As a result not only did the

Shukriya nomads fail to welcome the scheme as the government hoped and expected, but they are strongly opposed to it at least in the way in which it is at present operated.

The nomads have failed to identify with the scheme, as it demands the kind of revolutionary change in their life style that they do not wish to make, and for which they see no reason to make whilst profits from animal sales exceed those to be made from growing crops on the scheme. To them, development through irrigation means first and foremost water and food for themselves (that is, sorghum) and feed for their animals (that is, irrigated pasture and the grazing of crop remains after harvest). The scheme provides neither. So far they have taken up tenancies in order to have a claim to a place on the scheme if the system changes rather than to take advantage of the scheme as it is.

With the Nubians the problems are different. Here they were bribed by promises to agree to move to Khashm el Girba, and they feel that these were never fulfilled. Accomplished irrigation farmers from the Nile lands of the northern Sudan and wheat eaters, their major grumbles centre on having to grow cotton under an authoritarian management, on the failure to supply electricity for their homes, and on the provided houses whose design they dislike because it does not fit in with their traditions. They believe that given freedom to choose what crops to grow and to have some animals then they could make a considerably more viable enterprise of their tenancies.

Thus the basic problem in both cases centres on the different viewpoints and perceptions of the management and planners on the one hand and the tenants on the other. This was made even worse by trying to bring together into a single scheme a state owned sugarcane plantation of 10,000 hectares and peoples of such diverse traditions.

However, a further cause of difficulty has been merely to place the scheme on the river Atbara without considering the overall interrelationships of other schemes and existing agricultural systems. There is a clear need to integrate at least the major elements of the Butana (land between Atbara and Nile) economy into a coherent regional plan for Eastern Sudan.

The relevant items are :

(i) existing nomadic pastoralism;
(ii) existing peasant agriculture;
(iii) mechanised agricultural schemes in Gedaref and associated areas;
(iv) the partially developed Rahad irrigation scheme; and
(v) the Khashm el Girba scheme.

In the past each has been viewed as a separate self-contained entity.

Similar problems can be identified in the Gummuiya irrigation scheme sited on the left bank of the White Nile between Jebel Aulia and Omdurman. The scheme was first mooted and the basic investigations carried out in the 1930s as part of the programme to assist peasant cultivators whose traditional mode of life was upset by the building of the Jebel Aulia Dam. In the event the project was not proceeded with at the time, but was revived in the early 1970s. It was envisaged as a project to improve the availability of fruit and vegetables for the near 2 million people living in the capital conurbation (Khartoum, Khartoum North and Omdurman). In the event the scheme produces no fruit or vegetables and is a vast irrigated sorghum field with some berseem also produced.

Although some of the difficulties can be put down to maladministration, technical shotcomings and a dispute between peasants and government over land ownership, the real problem has been a difference in perception of development between planners and the people. The planners see the scheme as an abject failure whilst the peasants are less unhappy about it. The planners' viewpoint in development was the need to supply the demands of the market in the capital as they saw them, whereas the peasants viewed things differently. Situated on the desert margins with an average annual rainfall of less than 250 mm, they were used to crop failures, to sending their animals for grazing over long distances, and to supplementing their semi-nomadic way of life with seasonal work in the Gezira Scheme or in the capital. Their primary aim was to secure their own food supply and that of their animals. Sorghum was a crop that they have always cultivated during the rainy season for food; it can be cut green for animal feed; and it could be easily marketed in the capital. By contrast, fruit and vegetables were unfamiliar, required greater inputs and much more continued hard work. To these people development meant in order of priority a secure food supply, a secure feed supply, a relief to a hard life style, and only lastly an increased income. The planners saw increased income as the primary aim for these rural peasants.

In a sense the planners were unfortunate in the area they chose. In other parts of Khartoum Province such a scheme might very well have produced the anticipated results. In essence it was a sound scheme developed in the wrong place with the wrong people at the wrong time. The scheme may very well move in the planners' direction over the next ten years, but in its conception it turned out to be too "revolutionary" in its demands on the Gummuiya peasant.

At present there are some 160,000 hectares under officially approved mechanised agriculture schemes in Southern Kordofan Province mainly in the Habila area together with an unknown area under 'unofficial' mechanised agriculture farms. The programme began in 1968 as a government attempt to reduce the prevalence of food shortages in western Sudan. The programme followed the lines of similar development in eastern Sudan with farms ranging in size from 400 to 600 hectares (Davies, 1964). Alongside these schemes was another programme to modernise traditional peasant farming by bringing peasants together into co-operatives to work farms of 400 hectares. Seventy-three such schemes have been developed.

Both developments were conceived with little real assessment of the development potential of the area and without analysis of the regional development needs of the area. Thus, whilst the mechanised farms have eased the food problem by growing sorghum, the price has been the ruin of large areas through soil 'mining'. Wealthy entrepreneurs from town have stripped off the vegetation, cultivated the land to exhaustion and departed with their profits. Not only did local people derive little from the cultivation but they have lost valuable resources of timber and been left with exhausted and eroded land. The cleared timber was merely burnt on site as mere nuisance value to the entrepreneur desperate to catch the rains in his first season.

Sudan depends upon its wood resources for over 90 per cent of its fuel needs and vegetation is an indispensable part of house building, being extensively used for walls, roofs, doors and enclosures. Southern Kordofan is one of the main potential areas for wood supply, via the railway, for the Gezira and capital region of the country. Prices of firewood, charcoal and constructional timber have increased dramatically over the past five years in the capital in particular due to the increasing difficulty of procuring supplies. Inevitably under such conditions a great deal of illicit clearing of timber for firewood and charcoal burning has taken place without any programme of conservation and replanting. Furthermore, large areas under *Acacia senegal* and *Acacia seyal*, which both supply gum-arabic, have been indiscriminately cleared because *Acacia seyal* in particular provides excellent fuel and indicates land in good heart.

Although seventy-three modernised schemes are listed many of these are hardly functioning largely because the government's ideas of co-operation and those of the Nuba do not match and due to a dearth of good management. Nevertheless, all have cleared their land for cultivation and taken their toll of timber resources albeit they are by no means such a source of vegetational misuse as the mechanised schemes and were of course intended to directly contribute to the needs of local people.

Thus, as in eastern Sudan, a piecemeal uncoordinated sectoral approach sponsored by the central administration through its various ministries has had unfortunate consequences. It is plain that a coordinated regional plan for this area bringing together the various elements might have been a more successful form of development. The elements to be reconciled include:

(i) traditional cultivators;

(ii) nomadic pastoralists;
(iii) modernised farming;
(v) mechanised farming; and
(v) woodland resource management.

In the Sudan as in many other parts of Africa and the Third World centralised sectoral economic planning for development has not had the success to be expected from such large inputs of capital and effort. Schemes have not been as successful as they might have been because they were developed in isolation, with poor management and with little regard for the aspiration and needs of the local people as they see them. It is to be hoped that true regional planning will emerge through local initiatives under the new regional governments and that central government will not continue to write the regional development plans for them. Only truly integrated regional planning with the people can avoid a repetition of the misfortunes of the past and lead to a proper use of Sudan's natural resources.

REFERENCES

Abdel Wahab, A.R. (1976), Development planning in the Sudan, in *El-Hassan.*

EL-Arifi, S.A. (1978), *Local Government and Local Participation in Rural Development in the Sudan*, S.R.C. Monograph No. 10 (Khartoum).

Briggs, J.A. (1978), Farmers' responses to planned agricultural development in the Sudan (Gummuiya Scheme). *Trans. Inst. Brit. Geog.* Vol. 3, No. 4, pp. 464-475.

Davies, H.R.J. (1964), An agricultural revolution in the African Tropics : the development of mechanised agriculture on the clay plains of the Republic of Sudan, *Tijdschrift voor Econ.. en Soc. Geog.*, 53, pp. 101-108.

Davies. H.R.J. (in press) (Ed.), *Natural Resources and Rural Development in Arid Lands : Case Studies from the Sudan*, United Nations University (Tokyo).

Hanson, A.H. (1966), *The Process of Planning : A Study of India's Five-Year Plans, 1950-1964*, Royal Institute of International Affairs.

El-Hassan, A.M. (Ed.) (1976), *An Introduction to the Sudan Economy* (Khartoum).

Mirghani, A.R. (N.D.), The Ten-Year Plan of economic and social development 1961-62.—1970-71, *Sudan Economic and Financial*

Review, Special Issue No. 2.

Misra, R.P., Sundaram, K.V. and Prakasa Rao, V.L.S. (1974). *Regional Development Planning in India* (New Delhi).

Nimeri, S. (1977). *The Five-Year Plan (1970-75): Some Aspects of the Plan and its Performance*, S.R.C. Monograph No. 1, (Khartoum).

Prakasa Rao, V.L.S. and Bhat, L.S. (1964), A regional framework for resource development, *Bombay Geographical Magazine*, No. 1, pp. 35-50,

Sudan, Ministry of Culture and Information (N.D.), *The Six-Year Plan of Economic and Social Development, 1977-78 to 1982-83.*

Sudan, Republic of Khartoum (1961), *Ten-Year Plan of Economic and Social Development.*

Sundaram, K.V. (1977), *Urban and Regional Planning in India* (New Delhi).

Thimm, H.U. (1979), *Development Projects in the Sudan*, United Nations University (Tokyo), NRTS-1/UNUP-42.

8

INDEPENDENCE AND ECONOMIC DEVELOPMENT
A Case Study of Saint Lucia

FRIEDRICH STANG & PETER NAGEL

SAINT LUCIA is a small Caribbean island, which—having achieved political independence—is now striving for a more independent and balanced economy. This will be a difficult task since the island has apart from an agreeable climate and a beautiful landscape few natural resources. The present situation, with its historical and economic background, the plans for economic development and the problems connected are discussed in the following :

St. Lucia occupies a central position in the Windward Islands group of the Lesser Antilles at a latitude between 13°43′ and 14°07′N and a longitude between 60°53′ and 61°05′W. The island is 44 km long and 22 km wide and mainly mountainous. It has a tropical marine climate : the mean temperature range is between 24°C and 30°C with little variation during the year and between day and night. With about 2,000mm precipitation is abundant ; most of the rainfalls in the months from June to November.

Like India, St. Lucia is a developing country, but regarding size it is—like many of the latest newcomers among independent nations—right on the other end of the scale. With its 616 km² it could be compared to Pondicherry (479 km²); its population numbers only 1,22,000 (1981). In contrast to India,

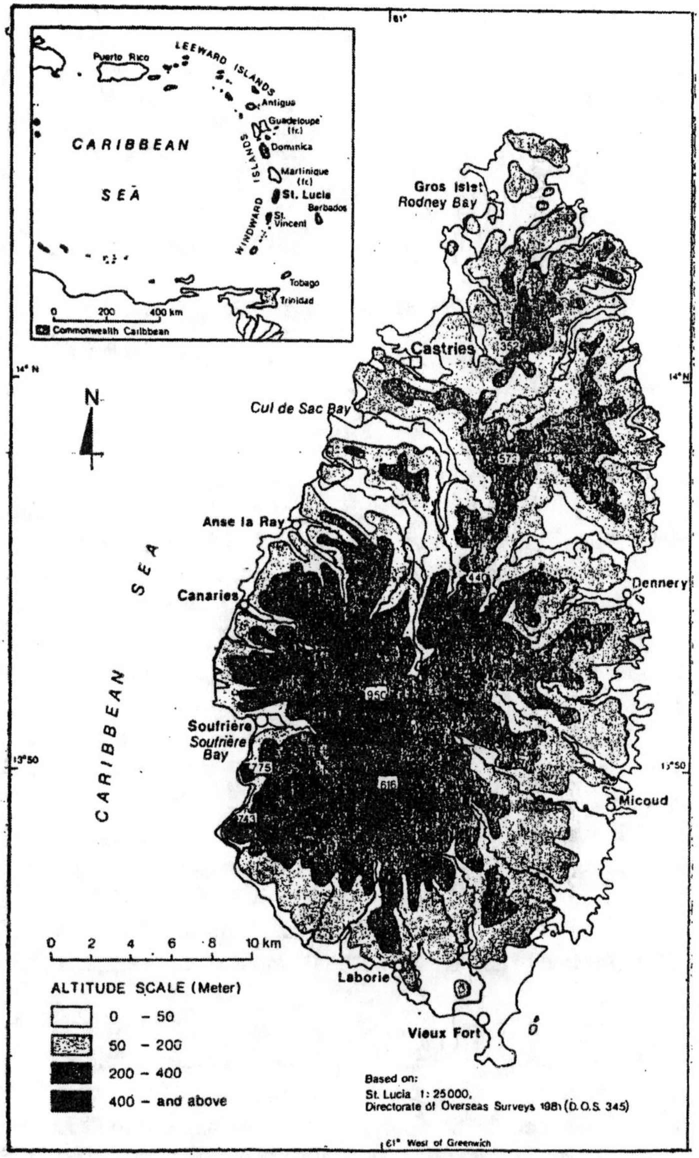

Fig. : 8.1 St. Lucia : Relief

where size, regional differences and the huge population of a subcontinent put a heavy burden on the government and institutions concerned with development, in St. Lucia it is possible for government officials to have an intimate knowledge of the people's problems and their local background. Smallness also has the advantage that even small investments can produce substantial rewards in terms of employment and per capita income.

On the other hand, the small size of the island, both in terms of population and resources, makes impossible all activities requiring economies of scale. The island is too small to constitute a protected market behind whose walls an industry could grow. On the international scale, its supplies to the world market are negligible, and its dependence on exports and imports will always make it vulnerable to any changes in the economies of countries far away. Thus, the problems involved with development are not only of a different scale to India's but also of a different kind.

An island of this size and population can hardly be expected to have an independent history. Apart from the early days with their subsistence economy, St. Lucia's political fate, just as that of other Caribbean islands, was determined by the power politics and the economic interests of European states. It is not certain whether Columbus, on his search for India—which he wanted to reach on a western course from Spain—discovered the island in 1498 or 1504. But when the Spaniards tried to occupy it, they met with fierce resistance from its then inhabitants, the Caribs, who came from continental South America about AD 800-1,000 occupying the Caribbean islands and conquering or enslaving the Arawaks; these were the first settlers, who had arrived on the islands between AD 200 and 400.

The Spaniards cared little for trade but were engaged in the search for gold which they could not find on the islands but only on the mainland; so the islands became important staging posts for their navy on the way there or back to Europe. They also afforded a hide-out for the many pirates—mainly of British origin—who tried to rob the Spanish ships of the gold which they, in turn, had plundered in South America.

Nevertheless the Spaniards firmly held on to most of the islands during the 16th century—and to some as late as 1890.

The interest of the later European powers—after the initial phase of piracy—concentrated on the production of sugar. In St. Lucia, it was the French who took over in the middle of the 17th century. Persistent quarrels with the British over the possession of the island, which changed hands 14 times, were finally decided by treaty to the advantage of the British. The French nevertheless left their traces. Place names on the island are French, and St. Lucians speak a vernacular 'patois', based mainly on French but including English words since English became the established language in 1842. Over 90 per cent of the population are Catholics, the religion introduced by the French. Perhaps the most important French legacy are the still existing laws of multiple inheritance. In many cases there are so many owners that nobody seems to know to whom the land really belongs. This causes bad land use by the peasants and gives them little incentive for improvements and investments. The present government is trying to transfer the title of the land to one person, but this is mainly for administrative purposes and will not bring a fundamental change in the near future.

After 1814, St. Lucia was a British colony governed from London. Executive and Legislative Councils were instituted in 1832. The first steps towards self-government were the election of three local politicians to the Legislative Council in 1924 and the establishment of trade unions. In 1951, Full Adult Franchise was granted and two political parties came into existence. In 1955, the offices of an Administrator, a Chief Attorney and a Minister of Finance were created. 1958 marked the birth of the West Indian Federation, comprising St. Lucia and the other British Windward Islands, as well as Jamaica, Trinidad, Barbados and the Leeward Islands. This confederation had lasted only four years when Jamaica and Trinidad withdrew. A new era began in 1967 : St. Lucia became a State with full internal self-government in association with Great Britain within the new confederation of the 'West Indian Associated States'; but foreign policy and defence were still responsibility of the British government. It was stipulated, however, that each of the states belonging to this association would be given full independence if parliament and the population voted in favour of it with a two-thirds majority. This they did and after 165 years of British rule St. Lucia became fully independent on 22 February 1979, remaining in the British Commonwealth.

THE COLONIAL ECONOMY AND SOCIAL CONSEQUENCES

Although political independence has been achieved, the economy of the island and even more the structure of its population is still strongly influenced by centuries of colonial rule. Apart from the Spaniards, the European nations who governed the Caribbean islands were mainly interested in the growing of sugarcane which started in the 17th century on the non-Spanish islands. In the Lesser Antilles sugarcane growing was first introduced in Barbados. In contrast to the other islands, Barbados has a flat topography which poses no difficulty to agriculture and transport. When demand grew in the 18th century, sugarcane also spread to the smaller islands further west. In St. Lucia, the first sugar plantations were started in 1965 and swiftly extended as the world demand for sugar rose.

There are two reasons for the late introduction. The established plantation-owners, the English in Barbados and the French in Martinique, were not interested in a new competitor and did everything to stop the development in St. Lucia. The second obstacle was the topography of the island. St. Lucia is made up of volcanic rocks and is very mountainous. Only the geologically older northern part of the island is flatter and denuded. Here the hills are not higher than 350 m, but in contrast to the rest of the island its north is rather dry. The central part of the island is dominated by steep hills which reach 950 m in the highest peaks; it is certainly not ideal for agricultural use. This central part is surrounded by broader, seawards-tilting terraces; these and the alluvial valleys of the rivers constitute the agricultural regions of the island. On the whole, the area available for sugar growing is small and transportation difficult.

The introduction of sugar plantation necessitated a large workforce for the harvesting and processing of sugar. It was "imported" as slave-labour from West Africa. For 1800, an estimated number of 14,000 negroes and 1,200 Europeans on St. Lucia is given by Jesse (p. 41). Disturbances and first attempts at a liberation of the slaves occurred during the French revolution. The abolition of slavery took place under British rule in 1838. Although the freed slaves had the right to acquire property, they did not have the means. The British had no interest in developing a small farmers' economy, but wanted to support the existing large estates which depended on cheap labour. Thus the freed slaves had no other option but to stay as labourers on the estates or to squat in the hilly land of the interior. In this way, an agrarian structure developed which was characterized by two extremes : a small number of large estates with their labour colonies, located on fertile soils and on flat land within easy reach of the ports, and on the other hand a great number of small farm-steads with a subsistence economy in the mountainous interior. But many negroes also settled on the hillsides surrounding the estates where they could work on their small plots and find additional employment so as to earn a little cash at harvest time on the estates.

Figure 8.2 puts the number of agricultural units in

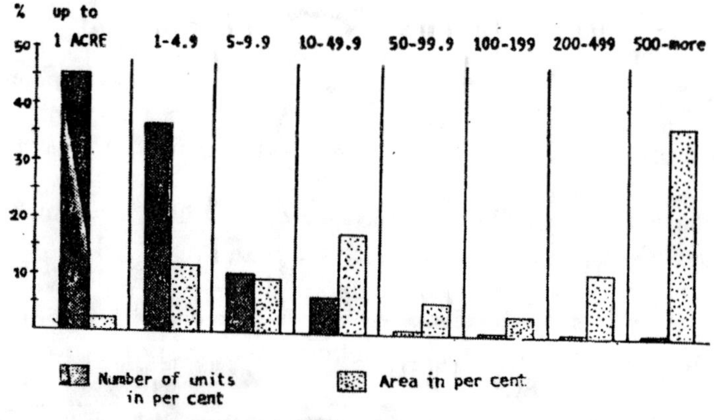

Fig. 8.2 : Size of holdings and areas percentage 1973
(*Source* : Government of St. Lucia, Annual Statistical Digest, Castries 1982, p. 14)

relation to area for 1973. It shows that the conditions have changed little since the first half of the 19th century.

It is the import of slave labour from West Africa from the 16th to the 18th century that explains the distribution of ethnic groups in the present population of the island : 90.3 per cent are negroes, 5.5 per cent mixed ("mulattos") and only 0.8 per cent white. There is a small minority of East Indians (3.2%) who came to the islands as plantation workers after the abolition of slavery—which leaves 0.2 per cent for "others".

Against this historical and social background, it is not surprising that working the land in general and particularly on the estates is not rated highly as concerns social prestige. It is this heritage that makes rural development difficult since it not only implies improved methods of working the land but even more a change of attitude.

From 1888 onward St. Lucia gained additional importance for the British as regular troops were stationed there. The building of barracks and other military installations offered St. Lucians jobs for the first time outside agriculture. Of even greater importance was the establishment of a bunkering station in Castries, a natural deep-water port on the leeside of the island. Between the 1880s and the 1920s, about 1,000 ships per year called in Castries to bunker coal which had been brought from Britain to the island. Considering that loading and unloading had to be done manually, the amount of employment thus created can easily be imagined. The availability of jobs in the port was also a strong pull-factor for the migration from the land to the town of Castries.

At this time the growth of Castries began, which is today the capital and has about half the island's population. Because the hills surrounding the town on three sides have restricted its growth, a suburban zone now extends along the sea towards the north. In this area two-thirds of the population and most of the non-agricultural activities are concentrated.

Due to the jobs offered in the coal trade, the percentage of those employed in agriculture decreased at an early stage.

Today it stands at 43.4 per cent (1981), which is rather low for a developing country. Having left the land St. Lucians began to look around for other employment, first in the other Caribbean islands and then from the fifties to the seventies in Great Britain resulting in a high emigration rate. New legislation in Britain and the independence of the other islands have meanwhile closed these outlets.

When coal lost its importance with the change from steam to diesel engines in ships, the resulting blow for St. Lucia's economy was softened by the fact that the United States leased in the north and south of the island land in the early forties to use for military installations. This created employment in the building industry and in road building. It also improved the island's infrastructure by supplying it with better roads and two airports. Agriculture was little affected by this development since the Americans retained their consumption habits and imported their food.

The sudden closure of the military bases after the Second World War caused mass unemployment and resulted in economic chaos. A fire in 1948 destroyed four-fifths of the town of Castries, but that turned out to be a blessing in disguise: financial aid from Britain helped with its reconstruction and provided new employment in the building trade. After the reconstruction period a new crisis set in which was partly overcome by the introduction of banana growing for export.

THE BANANA ECONOMY

Sugarcane growing had almost disappeared and by 1953 only one sugar factory was left working on St. Lucia. After unsuccessful attempts to introduce banana growing in the 1920s, the idea was again taken up in the 1950s and soon resulted in a change of the island's economy. Two English entrepreneurs, the Geest brothers, who ran a big horticultural and marketing firm in the UK, took the initiative. Backed by the British government they guaranteed to export to Great Britain all bananas of export quality produced in the Windward Islands. This guarantee involved a considerable risk since at that time the British market was dominated by bananas from South America and Jamaica. The Geest brothers bought estates on St. Lucia and Dominica on which sugarcane growing had been abandoned ; it was about 6,000 acres in St. Lucia. Under British management they started growing and exporting bananas. Other estate owners and small farmers followed suit.

Thus a new phase in the island's economic development began. Although the change from growing sugarcane to bananas appears to be a simple substitution of one agricultural export product for another, banana growing affected the whole structure of agriculture to a remarkable extent. In the days of sugar plantations, St. Lucia had two kinds of agricultural activity. On the one hand there were the large private estates producing for export and, on the other, there were the many small farmers who supplied the local market with agricultural produce for daily consumption and had an additional, very limited share in the export of cocoa, copra and spices. With bananas also the small farmers could produce for export. Bananas do not need processing in factories, and provided packing and transport are organized, the small farmer can offer them directly to the exporters. The growing time for bananas is short, cultivation does not need much labour and only little know-how. Harvesting takes place throughout the year and is not seasonal as is the case with sugarcane. So the work as well as a small income is spread more evenly over the entire year. Indeed, with the additional guarantee of a fixed price, bananas became a cash crop *par excellence.* But because of these advantages, the growing of food for local demand was very much neglected. In the course of a few years, almost all available land was taken over for banana growing. In 1964, sugar export had ceased and the export of cocoa and copra had dwindled to negligible quantities. St. Lucia had substituted dependence on one agricultural export product for another, which, however, offered considerable advantages to the small farmer.

The importance of an efficient organization has already been mentioned. Because the fruits deteriorate very quickly, the cutting, transport and loading of the bananas on ships

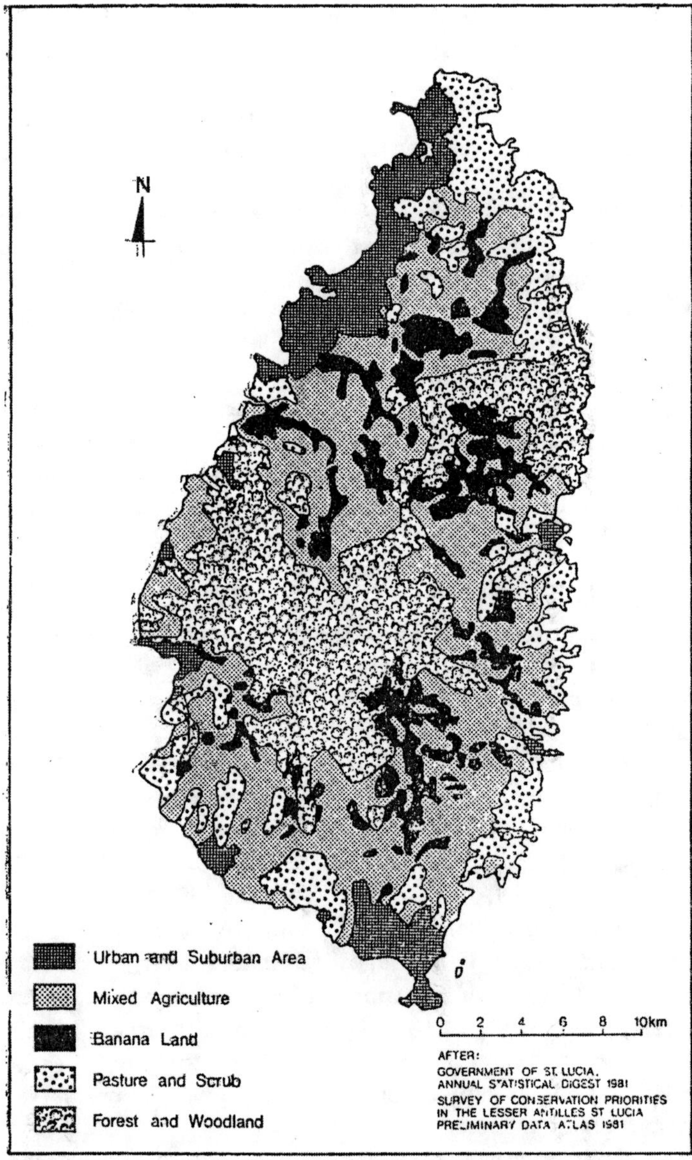

Fig. 8.3 St. Lucia : Land utilization

must take place within 36 hours. To make this possible, all banana producers, estates as well as small farms, are members of a cooperative, the 'St. Lucia Banana Growers' Association' (SLBGA), whose membership today is well over 10,000. The SLBGA coordinates the activities between farmer and exporter. Geest takes over in the port and pays the price to the SLBGA, which, in turn, pays the members of the co-operative. The rest of the money is used for the import of fertilizer, insecticides and to a considerable extent for transport on the island.

Transport is one of the main problems of banana growing in St. Lucia. From the fields of the small farmers, the bananas are first carried on the head to the nearest road, where the SLBGA has its collecting points. Previously the bananas were carried in bunches, which often caused damage through pressure when the fruits were lifted up or put down or while balancing them on the head. Therefore, the cooperative convinced the farmers of the need to cut off the banana bunches in the field and then carry the bananas in boxes or plastic containers, called pallettes, to reduce damage to the fruits. Transport from the collecting points to the washing and packing stations is then taken over by the cooperative's lorries. Because of the very poor condition of some of the roads, additional damage to the fruit may be caused. Road building and keeping the roads in good condition is, therefore, one of the main demands of banana growers and has high priority in the national development plan. Existing roads were straightened and asphalted, and, after 1975, 40 new roads were constructed, including feeder roads in the interior, bringing the total length of metalled roads to 314 miles, and of gravel and earth roads to 157 miles. They are mainly financed by development grants, which, however, do not seem to cover maintenance, so damage frequently caused by tropical rainfall is not repaired for some time. The feeder roads have opened up areas of the island which previously could only be reached by foot. This has led to more tree cutting and soil erosion and so has adversely affected the water storage capacity.

The estates are better off because they use cableways

or the old narrow gauge railways, that date back to the time of sugarcane growing, to bring the bananas directly to the washing and packing stations. This not only saves labour but also means a more careful handling of the fruit. From the washing and packing stations the bananas are transported to the ports of Castries or Vieux Fort where they are loaded and shipped to the U.K. once a week. For the last ten years, the fruits have been washed in desinfectant at the packing stations for better preservation and packed in cardboard boxes. This has several advantages to the transport of whole bunches. By not transporting the bunches there is about 10 per cent less in weight, so the ships can take more bananas and their loading time is reduced; the bunches and leaves are used on the island as fertilizer. Previously the ripping of fruits from bunches as well as their packing took place in Britain, where wages are much higher. Now the packing offers additional employment in St. Lucia. In addition, since 1973, the card-board boxes for packing are manufactured in St. Lucia.

Sea-transport is made more economical by loading all kinds of consumer goods, fertilizer, machinery etc., on the return journey from Great Britain. Apart from the rise in transport costs due to higher oil prices over the last ten years, the established system of two-way traffic between Britain and the Windward Islands in recent years appears to be suffering from the fact that a large amount of imported manufactured goods are of Japanese or American origin.

As a fruit, bananas are a relative newcomer on the world market. They could only find a market in the industrial countries with the rise in living standards, the use of refrigerator, ships and the setting up of a complex organization for growing, quick transport and marketing of the fruit since the end of the last century. The organizational problem and the investment needed led to the establishment of big companies which not only own the plantations but also the ships and who undertake the marketing. It is obvious that a concentration on big estates has advantages, and the companies opened up such estates in Central and South America.

For suppliers of very limited quantities like St. Lucia and the other Windward Islands banana growing and exporting only became feasible after the arrival of 'Geest Industries' as a big exporting and fruit trading firm in the 50s and after the British government had given preference to the Caribbean Islands' products. Supplies from other places than the Windward Islands and Jamaica are allowed into the British market only if the Caribbean producers cannot meet the demand. As Table 8.1 shows, this has happened in a number of years.

TABLE 8.1: **British banana imports from 1960 to 1981**

Year	*British banana imports in 1000 t*	*Market shares*		*in %*
		Jamaica	*Windward Islands*	*Others*
1960	300	42.3	19.7	38.0
1961	339	37.2	28.0	34.8
1962	343	39.4	29.7	30.9
1963	333	41.4	34.8	23.7
1964	331	44.4	39.0	16.6
1965	345	49.0	45.6	5.4
1966	344	52.9	43.6	3.5
1967	335	53.1	44.5	2.4
1968	330	46.1	50.3	3.6
1969	341	43.7	53.4	2.9
1970	325	41.8	52.8	5.4
1971	314	38.9	38.5	22.6
1972	311	39.9	36.9	23.8
1973	308	36.4	31.2	32.4
1974	306	22.2	35.6	42.2
1975	307	22.1	30.0	47.9
1976	313	23.6	37.1	39.3
1977	305	26.2	36.1	37.7
1978	315	22.2	41.3	36.5
1979	304	21.4	33.2	45.4
1980	322	10.2	22.4	67.4
1981	332	5.4	34.6	60.0

Source : FAO Production Yearbooks.

In other words, in spite of the stagnation in the banana market, the Windward Islands could sell more bananas in the U.K. than they actually produce. This is particularly so in those years when natural disasters have affected the production. A period of insufficient rainfall from 1971 to 1975 reduced crop yields, especially from the small farmers who have no irrigation at their disposal. Again in the years 1979 and 1980, hurricanes damaged a substantial part of the crop so that from August 1980 to April 1981, St. Lucia could not export any bananas. This clearly reveals the danger of having only one export commodity.

The U.K. is the only market for Windward Islands bananas, so the price is dictated by this market. Since the elasticity of demand is high, a rise in banana prices will lead to substitution by other fruits. Therefore, higher transportation costs caused by higher oil prices had to be borne by the producers. Fortunately, the island's preferential treatment was continued when the U.K. joined the EEC. This is of the greatest importance since in an open market the Windward Islands would hardly be competitive against the big estates in South and Central America.

THE POST-INDEPENDENCE ECONOMY

To offset the dependence on banana exports, the Island's Development Plan is trying to achieve a diversification of the economy by engaging in additional activities. First it seeks to foster the tourist industry which will not only offer additional employment but will also have a multiplier effect on the rest of the economy. Secondly, it envisages industries for the local market aiming at import substitution, and thirdly, it wants to promote industries located in Free Zones producing exportable goods. High priority is also given to the development of agriculture. The idea is to increase output and diversification, which would not only result in additional products for export but would also supply locally needed basic food stuffs.

Tourism

Of all the Government's attempts to put St. Lucia's economy on a broader base, the development of tourism has been the most successful. A number of factors have helped this development. Higher incomes and longer holidays in the industrial countries have led to an increase in tourism world-wide. With growing incomes spending on travel has risen at a faster rate than other expenditures.

In the initial stages, the Caribbean Islands mainly had tourists from the USA and Canada who were attracted by the warm climate during the winter months. At that time only a few islands had good air or sea connections to the mainland. When after 1955, Cuba dropped from the list of holiday places, tourists went to Puerto Rico and Jamaica instead. Of the Lesser Antilles, the first to attract tourists were the flat coral islands of Antigua and Barbados which, compared to the volcanic islands, had better and whiter sandy beaches and good roads. Before the introduction of charter flight travel only wealthy people could afford a holiday on one of the Caribbean islands, and, because of this, the hotels are still of the luxury class.

It was in the middle of the 1960s that tourism also reached the smaller Windward Islands. A touristic infrastructure was almost non-existent and places like St. Lucia were advertised as exotic untouched paradises with lonely beaches and a friendly, hospital population. From 1969 to 1973, hotels were erected by international hotel groups and the number of available hotel beds quadrupled to almost 3,000 in 1981. The foreign hotel companies were allowed a tax holiday and tax-free imports of building material and equipment for the hotels. On the government side, great efforts had to be made to guarantee supplies of drinking water and electricity. The number of visitors rose at a remarkable rate : In 1970, 30,000 visitors were recorded, and, in 1981, 71,000 visitors. The airport, which was built in the 1940s by the US in the extreme south of the island, was reopened again as 'Hewenorra International Airport'. Aircraft of any size can land here whereas the small airport near Castries is only suitable for small planes which connect St. Lucia with the other islands in the Caribbean. When the age of charter travel began, the international airport gave St. Lucia a big advantage over other islands in the Windward Group, because it could be reached directly. Most of the hotels are in the north-west of the island, so the existing small road from the international airport to Castries and further on to the north was improved and the transfer from the airport to the hotels, which had previously taken three hours, was reduced by half. This was particularly important for travellers from Europe who had already sat through a ten hours' flight.

In the north a holiday centre was created at Rodney Bay. This project was started in 1970 by the Government of St. Lucia in co-operation with the Commonwealth Development Corporation and private parties. In the first stage, Pigeon Island was connected to the mainland by a dam and 70 acres along a beach of fine sand reclaimed for hotels. The second stage of the project consisted in reclaiming the mangrove area near the fishing village of Gros-Islet. This area was partly drained so as to provide building land for hotels and holiday colonies and partly dredged to make a 80 acres artificial lagoon for a protected yacht-harbour. The north-west coast from Castries to Cap Estate including Rodney Bay will be the main centre for tourist activities but two more areas have been marked out: the south with Vieux Fort, which already has a big hotel, and the Soufriere area, the most beautiful part of the island, which still has to be developed.

Apart from the tourists who spend their holiday in a hotel or a "villa", three other groups are of importance. Visitors on cruise ships usually only stay a day in Castries and may take an excursion into the interior. At the other end of the scale are the long-term "tourists", who come to St. Lucia to buy or build a house and retire here. A third group are yachtmen who visit St. Lucia with their own yacht or for a charter-holiday.

Most of the visitors from cruise-liners come during the winter. Many shipping companies concentrate on the Caribbean at this time of the year. The climate is warm and distances from one port of call to the next are short. The port of Castries has a long tradition as a port of call for American cruise-liners ; it is intended to start the construction of a new cruiser terminal with shops and other facilities this year.

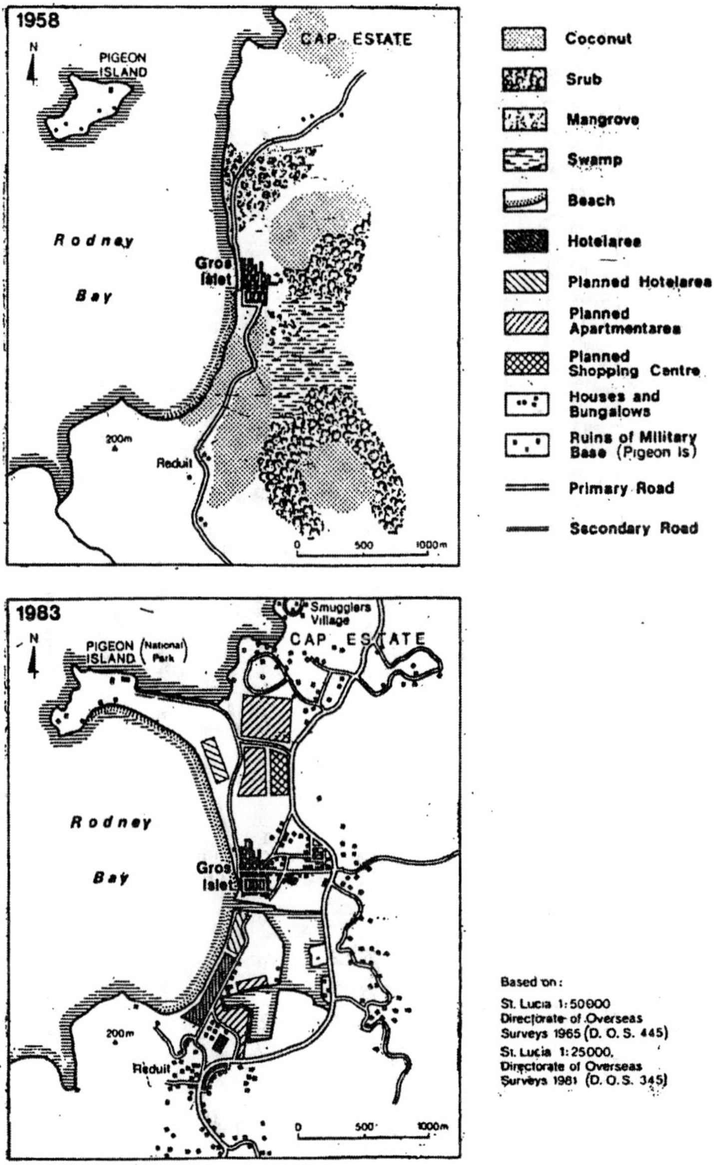

Fig. 8.4 Tourism development: Rodney Bay

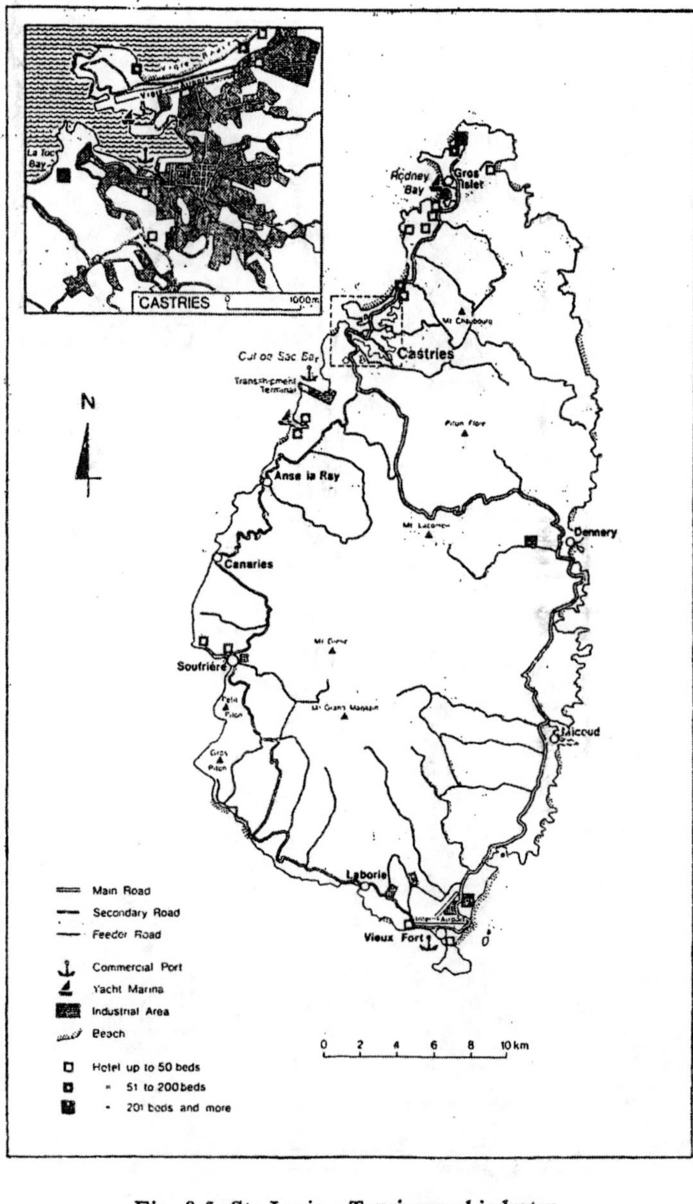

Fig. 8.5. St. Lucia : Tourism and industry

The extreme north-west is the area reserved for foreigners who want to retire in St. Lucia. This Cap Estate was used in colonial times for sugar and cotton growing and later—because it was rather dry—for cattle raising. It was parcelled out in the 1970s and the government has introduced legislation to enable foreigners to acquire land.

Three marinas are available for yachtmen : Rodney Bay lagoon, a part of the port of Castries, and Marigot Bay, about 12 km south of the capital. For yachting, St. Lucia has two advantages : because of its international airport cheap and easy flights are available which also make it a good location for the charter business hiring-out yachts. There are two big charter firms located on St. Lucia. The other advantage of St. Lucia are the many small islands of the Grenadines group between St. Vincent and Grenada, hundreds of mostly uninhabited islands within easy reach, a paradise for sailors.

Table 8.2 shows the development of tourism and the origin of the tourists.

TABLE 8.2 : **Number of tourists in 1967 and 1981**

	1967		*1981*	
	Number of tourists	%	Number of tourists	%
Caribbean	7,589	46.9	16,739	23.7
USA	4,461	27.6	12,352	17.5
Canada	1,394	8.6	12,094	17.2
South America	340	2.1	1,797	2.5
United Kingdom	2,067	12.8	14,373	20.3
Other European Countries	218	1.3	12,753	18.0
Rest of the world	123	0.7	588	0.8
	16,192	100.0	70,696	100.0

Source : Government of St. Lucia, Annual Migration and Tourist Statistics 1982, Castries, 1983, p. 7.

The rising number is mainly due to the availability of cheaper charter flights to the international airport. In spite of the fact that most of the tourists come on package-tours,

St. Lucia has not been taken over by mass tourism. The hotels are few in number, and hidden behind palm trees, so they do not spoil the natural beauty of the island.

Figure 8.2 shows that apart from the visitors from the Caribbean area—of whom only few can be regarded as "tourists"—North Americans are first on the list. Second place is taken by British visitors. The number of German visitors has grown considerably in the last years, since there is a weekly connection from Germany to St. Lucia and a German hotel is available in the north of the island. There are also direct charter flights from London, Montreal and New York. The average duration of stay has dropped from 14 days in 1970 to 8 days in 1978. Most Europeans come for at least a fortnight, but Americans hardly stay in one place longer than a week and better transport, together with special tickets for visiting several Caribbean islands, has made island-hopping possible and attractive.

One of the problems of tourism, its seasonal character, has been overcome by the charter business, which spreads tourist arrivals more evenly over the year ; it is mostly Americans during the winter in search of a warmer climate and mostly Europeans in the summer in search for the sun.

The development of tourism has had various effects on the economy of St. Lucia. The erection of hotels and also of private houses led to a building boom which provided employment for a number of years. Income from tourism has certainly had a stimulating effect on the economy of the island but it has also attributed to a higher inflation rate. The hotels' profits are usually transferred back to foreign countries and in spite of the tax holiday limit of six or ten years, most of the older hotels have been successful in transferring ownership or setting up new companies to avoid taxation and extend their tax holidays.

One can say that the main importance of tourism lies in the employment it offers. In 1978, 2,300 people, mainly women, were employed in hotels and restaurants, and another 4,400 people in sectors connected with tourism, from taxis to cottage industries, which supply souvenirs for the tourists. Today, tourism takes third place as employer behind agriculture and administration. Income from tourism takes second place after the export of bananas. The high unemployment rate and the demand for "clean" jobs in the tourist trade, which are much preferred to working on the land, has made it possible for the big hotels to hire labour at very low wages.

Industry

The opportunities for the setting up of industries are limited, since St. Lucia has no raw materials for industrialization except for agro-based industries. In the 1950s a co-operative was founded to buy copra from the coconut producing farmers and process it to edible oil, margarine and soap. The hurricane of 1980 considerably damaged the coconut trees which take a longer time to grow than the bananas and as a result the factory is not working at full capacity. Nevertheless, this enterprise is the biggest non-agricultural employer in the south of St. Lucia. The production of rum now-a-days has to rely on imported raw material because cane growing has disappeared from the island.

The government hopes to foster industrial activity by establishing two industrial estates, one in Castries and another in Vieux Fort, where industrial sheds are being built and let to entrepreneurs. More industrial estates are planned in the little towns of Dennery and Soufriere. The existing industrial estate at Bise, north of Castries, is meant for small local firms. Vieux Fort offers more space. It has a firm producing the boxes needed for exporting bananas, not only for St. Lucia but also for the neighbouring islands, and a brewery under Dutch management which supplies the Caribbean area including the French islands. Both firms employ more than 200 people but they are more capital- than labour-intensive. Their location in St. Lucia is primarily due to the good transport facilities. For most industries aiming at import substitution the local market is too small, with the exception of beer and soft drinks. Cement blocks are manufactured from imported cement with the addition of local raw materials like sand and gravel.

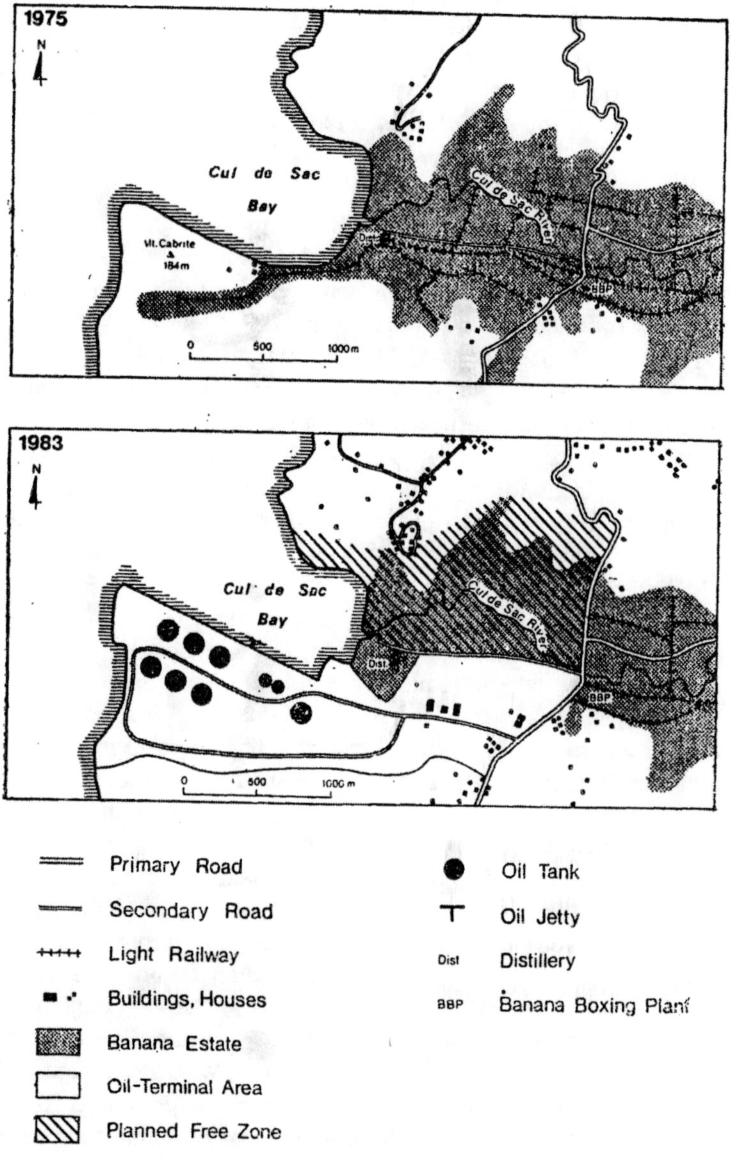

Fig. 8.6 : Oil-terminal development : Cul de Sac

On a quite different scale from the industrial estates is an oil transshipment port built after 1977 by the American firm of 'Amerada Hess Corp' in Cul de Sac Bay. The firm owns the biggest refinery in the western hemisphere in St. Croix on the US Virgin Islands and other refineries in the USA. The oil is transported to the transit shipment terminal in St. Lucia, partly stored there and partly transferred to smaller tankers which take it to the above refineries. A big refinery was also planned in Cul de Sac but because of the reduced demand for oil, the construction was postponed.

The location at Cul de Sac Bay was chosen because, after dredging, the Corporation was able to create a harbour with a water depth of 92 ft. This depth will permit virtually any very large crude carrier in the world to discharge its entire cargo at the facility without first being lightered. After a hill was cut down from an elevation of 485 ft to 120 ft, construction of storage tanks with a capacity of about six million barrels followed. Inbound vessels pump cargo into the elevated tankage, and outbound vessels are loaded by gravity feed. During construction work about 400 local labourers found employment there. But now-a-days the few jobs available are held by foreign specialists. The St. Lucia government levies a tax of 0.02 US dollar/barrel for the transshipment of oil in this port. With the projected 100 million barrels the income from this would amount to 2 million dollars.

St. Lucia can offer cheap labour in abundance, although mainly unskilled. About 30 per cent of the working population is unemployed. The government is, therefore, trying to follow the example of other islands in the Caribbean, particularly Puerto Rico, in establishing "Free Zones" which would process imported raw materials and re-export them to industrial countries. There are plans for a Free Zone at Cul de Sac Bay. One firm in the electronics branch is operating at Vieux Fort Industrial Estate making use of the available labour and tax concessions.

Another advantage of St. Lucia lies in the good traffic connections to other islands and the American Continent. There are two easily accessible, well equipped ports and two airports.

The port of Castries has facilities for handling containers and it is hoped that the central position of the island among the Windward and Leeward islands will help to make it a distributing centre for container traffic.

PROBLEMS OF A VERY SMALL COUNTRY

Since 1979 St. Lucia has been an independent sovereign state. But it is still a long way from economic independence, and it is doubtful whether, or to what extent, this can ever be achieved. Among developing countries St. Lucia is not too badly off with a per capita Gross Domestic Product of almost 1,000 US dollars in 1981. But the constraints of economic growth affecting most tropical countries are aggravated by its smallness and isolation as an island.

Among the natural handicaps affecting small and mountainous islands are soil erosion and water shortages due to the small catchment areas available. They become more important with economic development and the resulting pressure on the land. Droughts or catastrophes like a hurricane would in mainland countries or larger islands only be felt in part of the territory whereas in St. Lucia the entire island is affected and no part can make up for losses in other areas.

The burden of demands made by the administration and foreign representation is out of all proportion in a very small country although many government employees such as teachers or nurses are required on a per capita basis. Government and social services in St. Lucia employ 22 per cent of the working population and originate with 34.2 million dollars ECC (East Caribbean currency) the largest share of the GDP (1981), higher than agriculture (31 million dollars ECC) in which 43 per cent of the working population are employed !

Considering that one of the cheapest means of transporting goods is by sea, islands would appear to be at an advantage. It is only when one considers that the economics of sea transport depend on large quantities of goods shipped to one destination that the smallness of the island explains the high cost of imports, which not only make life very expensive for the local people but also for the tourists.

The diseconomics of scale of an industrial production for a small market like St. Lucia has prevented the establishment of industries. Moreover, the capital needed is hardly available in St. Lucia. As is common in plantation economies little of the money made from sugar was ploughed back as an investment in the island. And most of the money sent by St. Lucians working in Great Britain is spent on supporting the members of the family who stayed at home.

If the local market cannot support industries, setting up Free Zones with foreign capital, producing for the world market, will not be easy either. Tax concessions and plenty of labour are available and an investor would certainly be granted additional special concessions to overcome special difficulties. But these are bad times for starting new ventures. There are already a number of established and flourishing Free Zones in the Caribbean with better infrastructure and there are other places in the world where the workforce could be more efficient.

Taking into account economics of scale, the concentration on the banana industry is certainly an answer to the problem of limitations, St. Lucia is up against as a small island. Thanks to the arrangements made by Geest and the preferential treatment by Great Britain, it can supply a product in sufficient quantities to keep the cost of handling, transportation and marketing within reasonable limits. Thus, there is a strong case against diversification into other agricultural export products.

On the other hand, concentration on banana growing leaves the island wide open to fluctuations on the markets or in exchange rates. St. Lucia's dependence can best be shown by the fact that for St. Lucia bananas constitute one third of its exports (in value), whereas St. Lucia's share of the world market is less than 1 per cent.

But also regarding the production of food for the home market it seems difficult to convince farmers of the need for diversification. Banana growing has supplied them with a small but steady and guaranteed income throughout the year and given them a chance to exist alongside the big estates. This has made them reluctant to invest labour and capital for other productions, bananas giving much easier and better returns. Consequently, St. Lucia's imports are rising not only to meet the daily needs of the local market but even more the demands of the hotels whose visitors are not used to the heavy starchy vegetables like yams or sweet potatoes grown in St. Lucia.

The great advantage of tourism, compared with other economic activities, lies in the fact that the small size of the island is no serious limitation as long as sufficient number of flights are available to make it easy to reach. Here economies of scale are to be measured against quality, mass tourism against the quieter holiday. That this type of holiday is more expensive than most has not yet had an adverse affect on the development of St. Lucia's tourist trade. It may be that wealthier tourists are less hurt by economic recessions or by poor exchange rates. St. Lucia has weathered the recent recession rather well and its main setback was caused by the hurricane of 1980.

If the disadvantages of being a very small island generally outweigh the advantages could this situation be overcome by the creation of a customs union or a political federation? The attempt made in 1958 to set up a West Indies Federation failed, because no agreement could be reached on the kind of federation, on a customs union, common revenues or on major governmental departments on a federal basis. There are several reasons for the difficulties facing the establishment of a federation. O'Loughlin (p. 5) points out that a federation as a form of government has not worked well in emerging countries in the Third World. For a long time they had been governed from outside. Now that they have finally achieved their independence, they are reluctant to give up part of their sovereignty to a supranational authority. There are also no strong bonds between the islands, except that they were and are still united under the British Crown. Each one's economy and administration were closely connected to Great Britain. In comparison to this orientation the linkages between the islands were of small importance.

If, therefore, a federation with a strong central government is hard to imagine, it is also unlikely that a customs union could result in a breakthrough. All the islands united would still represent a small market and transport costs between the islands would reduce the advantages of being a single unit. Only a few industries would be attracted by such a small market. And which island will have these industries? There can be little doubt that the more developed islands would stand to gain more than the less developed and regional disparities would widen further.

There seems to be no easy answer to the question of how to achieve progress under present conditions? Small may be beautiful but it raises a lot of problems—and it can be rather expensive.

REFERENCES

Beckford, G., (1978). *Caribbean Economic*, Kingston.

Blume, H., (1968). *Die Westindischen Inseln*, Braunschweig.

Brock, C , (1976). *Saint Lucia*, London.

Caribbean Conservation Association, St. Lucia, (1980). *Preliminary Data Atlas*, Castries.

Chernick, S., (1978). *The Commonwealth Caribbean*, World Bank Publication No. 16. Washington.

Demas, W., (1965). *The Economics of Development in Small Countries with Special Reference to the Caribbean*, Montreal.

Government of St. Lucia; Statistical Office, (1982). *Annual Migration and Tourism Statistics*.

Government of St. Lucaia, Castries (1983).

Government of St. Lucia (1982). *Annual Statistical Digest 1981*, Castries.

Government of St. Lucia (1981). *Statistical Pocket Digest*, Castries.

Government of St. Lucia, (1979). *St. Lucia at Independence*, Castries.

Government of St. Lucia(1977). *National Plan*, Castries.

Haas, H., (1976). *Die Industrialisierungsbestrebungen auf den westindischen Inseln. Tübinger Geographische Studien*, 68. Tübingen.

Jesse, C., (1964). *Outlines of St. Lucia's History*, Castries.

Le Franc, E. (1980). St. Lucia. In : Caribbean Development Bank (Publ.) : *Small Farming in the Caribbean*, Barbados.

Lowenthal, D. (1972). *West Indian Societies*, London.

O' Laughlin, C. (1968) *Economic and Political Change in the Leeward and Windward Islands*, New Haven.

Romalis, C. (1969) *Barbados and St. Lucia : A Comparative Analysis of Social and Economic Development in two British West Indian Islands*, Ph D Washington University.

ULG Consultants Ltd. (1976). *The Windward Island Banana Industry, Inputs and Services*, London (unpublished).

United Nations Development Programme (1975). *Technical Supplement to the Draft Advisory National Physical Development Plan for St. Lucia*, Castries (unpublished).

World Bank. (1979) *Current Economic Positions and Prospects of St. Lucia*, Washington.

9

A COMBINATORIAL STUDY OF COMPLEXITY OF DRYLAND FARMING IN THE DECCAN : INDIA

GRAHAM P. CHAPMAN AND BARBARA HARRISS

DRY FARMING is renowned for its complexity. In ICRISAT's[1] six 'laboratory villages the average number of sole crops and intercrops is 63 ; in the AICRPDA[2] sample of 261 producers used as the basis for this chapter there are 62 crops grown in monoculture, or in intercrop combinations, of which there are 99. Some of these intercrop combinations have up to 5 crops growing together.

Such intercropping practices have often been ascribed to risk-spreading in areas of uncertain rainfall, but it is also clear that they have significant benefits in inhibiting the diffusion of diseases, complementing the demands on soil nutrients and even replacing some (through fixation of nitrogen) and also spreading labour demand and food production over longer time periods.

This study does not include any comments or analysis on such intercropping practices, but it will be readily apparent that the methodology used would be appropriate for such a study.

At the next scale above the fields comes the farm. Farms will be characterised by the combination of crops they produce, whether or not they are grown as single crops or intercrops. Farms, or more accurately farmers, will also be characterised by the extent to which they produce surpluses which are traded, and by the produce which they receive from others, either in payments in kind or as a result of cash purchases. This study limits itself to the 14 crops of the AICRPDA data set, which dominate production in the three regions, but is broad in its approach in that it does look at farmers in terms of the combinations of their involvement in the crop economy of the region : that is in both the combina-

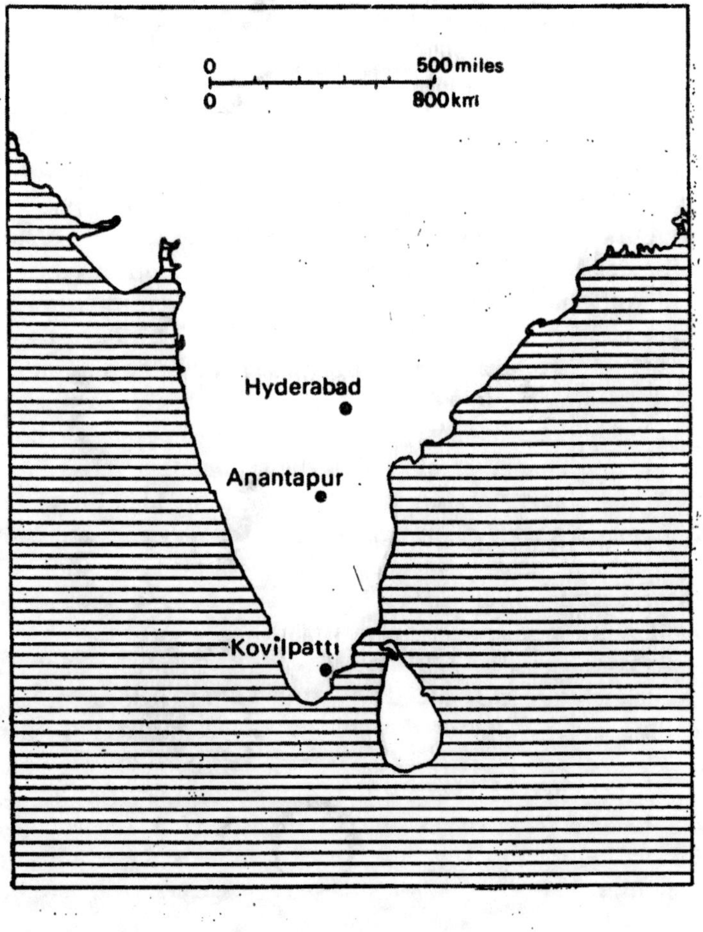

Fig. 9.1. Location Map.

tions of their farm production, and also in transactions of produce. The three regions are : Hyderabad, Anantapur and Kovilpatti.

The data set

Since Bangalore (a fourth region in the AICRPDA sample) did not provide data on purchases, 181 out of the initial 261 producers are covered by this analysis. They have been classified both by membership of one of the three regions and by membership of one of five social classes. These latter have been defined by the ratio of labour hired in to labour hired out ; $E = \frac{Li}{Lo}$ where *Li* is number of man days hired in, and *Lo* the number of man days hired out.

The classification is as follows :

Class	*Code*	*E value*
Marginal farmers	A	<0.07
Small farmers	B	0.071–.70
Middle farmers	C	.71–1.43
Rich farmers	D	1.44–14.3
Capitalist farmers	E	>14.3

Ten crops define the foodgrains economy : paddy/rice (expressed in paddy equivalent using a multiplier of 1.5), ragi, bajra, jowar, small millet, maize and wheat among cereals and redgram, cowpea and horsegram among pulses. Also, in this study, four important market-oriented food and agro-industrial crops are included : tomato, cluster bean, groundnut and cotton. Together the 14 crops account for 89 per cent of total production in the three regions.

Table 9.1 shows the distribution by social class of the sample. Table 9.2 shows the percentage distribution of each social class according to the number of crops grown by each farmer. It is noticeable that there is an increase in complexity (number of crops grown) with higher social status, also that with the two highest groups the distributions are bimodal. Analysis shows that these result from the fact that there are two types of farming, as noted above. Those with irrigated

TABLE 9.1 : **Characteristics of sample**

Social class	*Number*	*Percentage*	*% contribution to production*
Marginal farmers	36	20	5
Small farmers	70	39	16
Middle farmers	8	4	2
Rich farmers	33	18	23
Capitalist farmers	34	19	54

land grow paddy and some pulse, and represent the first peak in the distribution, whereas the second peak represents the farmers who grow complex combinations on dryland, with or without some irrigated land as well.

Market involvement

Farmers will obviously be involved in the economy of some crop if they grow it, but they may also be involved even if they do not—by purchase, or by receipt of payments in kind for rent, labour and debt. In turn these receipts may be used for further payment by kind or for sales. Unfortunately, it is not possible with the data we have to analyse all such activities. We are left with three categories only : produce, purchase and sell, so that the non-cash payments in kind are excluded from the explicit data. However, such payments are implied by some of the categories we use, as will become apparent below.

With the three categories we have, we could define eight kinds of market involvement for each farmer for each crop.

(1) uninvolved,
(2) produce only,
(3) sell only,
(4) purchase only,
(5) produce and sell,
(6) produce and purchase,
(7) sell and purchase,
(8) produce, sell and purchase.

It can be seen that with category 3 there must have been

TABLE 9.2 : **Number of crops grown by social class**

Social class	*Number of crops* 0	1	2	3	4	5	6	7	8	9	10	*Median no. of types of foodgrains*	*Total no. of cultivators*
A	16	30	25	12	2	9	2	—	2	—	—	1	43
B	3	20	29	18	13	9	6	1	—	—	—	2	95
C	0	10	35	30	10	15	—	—	—	—	—	3	20
D	0	13	22	15	25	13	11	12	—	—	—	4	56
E	0	8	22	9	8	15	19	8	4	—	—	5	47

Note : Figures are percentage distribution across the rows of each social class.

receipts in kind and the same will be true of categories 5, 7 and 8 if the sales exceed either the production or the production plus purchases.

How many possible combinations of market involvement are there with 14 crops ? For each of these 14 crops any of the above eight categories will be possible, so we have $8^{14}=4.4\times10^{12}$ possibilities, a number in the range of millions of millions. Another way of getting the same answer is to say that there are 14 crops and three possibilities for each crop—for example, produce, buy or sell, so that there are 14×3 'attributes' the farmer could have : for example, produce bajra, sell bajra, buy bajra, produce rice, sell rice, buy rice, etc. He may either have or not have each of these attributes : so we have $2^{14\times3}=2^{42}=8^{14}$ possibilities. We can immediately see why there is so much room for experimentation with cropping systems.

In this study we have simplified the problem by saying that a farmer is involved with a crop if he produces and/or buys and/or sells it. Then we have 2^{14} combinations (16,384) which are possible. But it must be borne in mind that when we say a farmer is in the 'bajra, rice, tomato' economy, for example, it does not necessarily mean he is producing all of these. He may be accepting payment in kind and then selling.

The next question is to see whether there is a method to discover what combinations exist in the data, and to see what story these combinations tell. We do this by using a new methodology known as Q-analysis, and some accompanying computer programmes.

A NOTE ON Q-ANALYSIS

Q-analysis offers a radically new methodology for describing complex systems. It embraces philosophical insights into the nature of data, conceptual rigour in the construction of generalising classification and analytical procedures that summarise the structure inherent in data matrices, without expecting unwarranted assumptions. Clearly, this is not the place to offer an introduction into the full scope of this methodology. Interested readers are referred to Atkin,[3]

Johnson,[4] Chapman,[5] and Gould *et al*[6]. Some guidelines are necessary to grasp what follows.

Q-analysis suggests that there are structures in the world which influence the nature of activity that can take place on the structures. In saying this it is not dissimilar from Marxist theories which insist that the relations of production can influence the nature of economic activity. However, it does not theorise on what the structure might be, but provides an algebraic language for describing them. In so doing it insists on a distinction between backcloth (the structure) and traffic (the data that live on the structure). In the case of farming, structures might be defined in terms of a farmer's possession of land, capital, technical knowledge, marketing access, and specific kinds of seeds, etc. The traffic in this case might be the total outputs of the different crops. None of this sounds particularly revolutionary, but some of the differences which the approach brings will be apparent at the next step. In most normal analyses data would be collected in a single matrix, or possibly two matrices, with the individual farmers as the rows, and the columns containing the data on backcloth elements and the resultant yields. Reduction of these data would be made by multiple regression, which would only handle a single dependent variable at a time, and which would assume that the independent variables were also independent of each other, and contributing in a linear additive manner. Q-analysis insists that the backcloth for each farmer be constructed as the total of all his elements. His traffic (yields) is then related to this holistic totality. Two farmers who have the following elements B_1,B_2,B_3,B_4, and B_1, B_2, B_3, B_5, are treated as completely distinct, even though they have some elements in common. Even factor analysis operates on the assumption that the variables can be summarised by virtue of their variation over all observations, and then reconstitutes the observations in terms of their loadings on the factors. In other words, for any one individual, the unity of the variables, their unique assemblage for that individual, is completely ignored.

Since a farmer is represented by all his backcloth elements simultaneously, it is possible to conceive of him as a

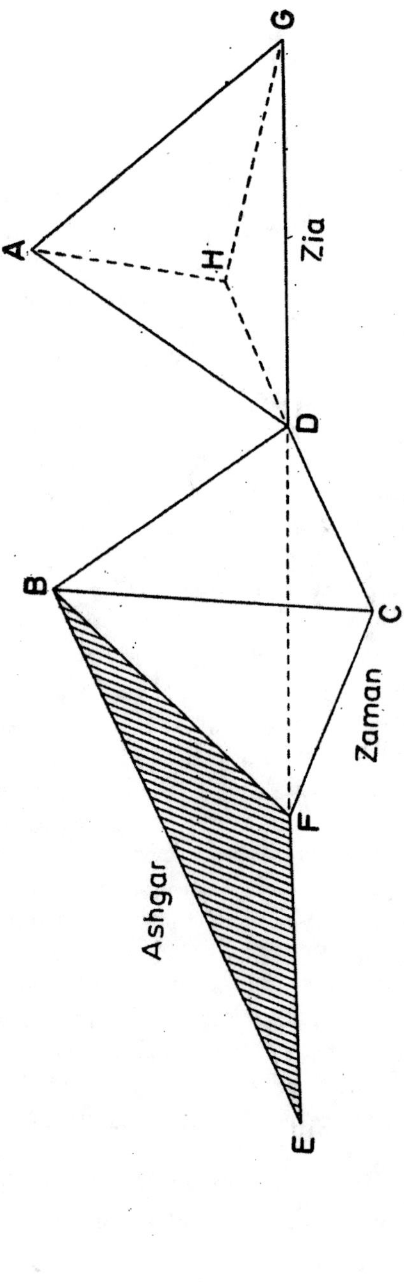

Fig. 9.2. *Part* of the complex of Fig. 9.3

polyhedron in multidimensional space. The polyhedron representing one farmer may share certain faces with the polyhedron representing another farmer (Fig. 9.2). In a complete data set, the polyhedra representing all the farmers may be strung together. There is no way that one can 'see' such a multidimensional complex, but there are ways of representing it so that we can appreciate some of its structure. Starting with the dimension of the highest dimensional polyhedra, an algorithm in Q-analysis searches the structure at successively lower dimensions to see which polyhedra are connected at which dimensional levels. In Figs. 9.3 and 9.4 the connectivity structure of five farmers and eight crops is shown. The traffic is expected to be affected by these dimensional levels. Thus it would not be possible for a deal involving five crops to be struck between farmers Taher and Abul. Crudely put, the structure which Q-analysis reveals puts farmers into groups which might exhibit variation from one end of the group to the other, but in which along a chain of connection individuals overlap with others in many respects. What then becomes of significance is the breaks in the chain. Such breaks inhibit transmission of traffic in the structure. In the analysis which follows in this chapter, the connectivity patterns have been displayed using graphic output direct from the computer. Alongside the stars representing the number of vertices of each polyhedron are the names of the crops involved for

Farmers	Crops							
Abul	A		C	D		F		H
Ashgar		B			E	F		
Taher	A	B			E		G	H
Zia	A			D			G	H
Zamen		B	C	D		F		

Fig. 9.3 A relation between farmers and crops

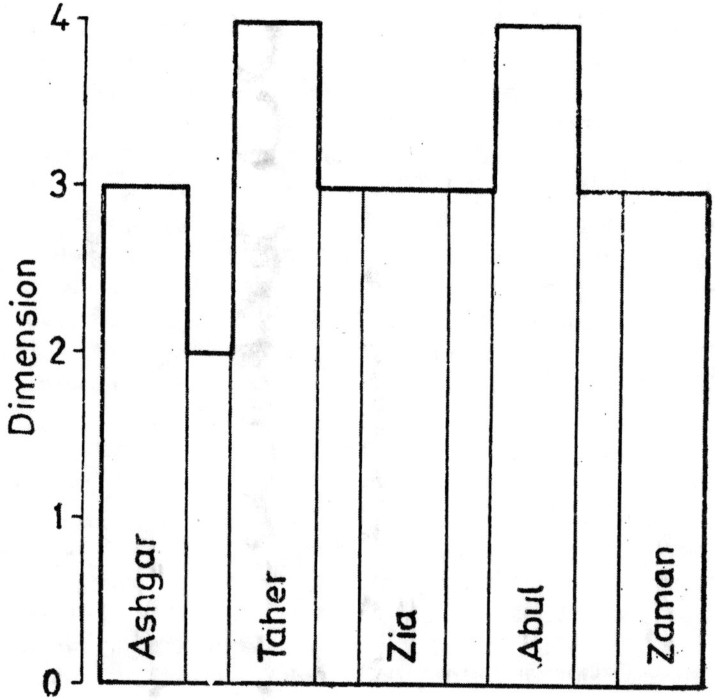

Fig. 9.4 The connectivity of Fig. 9.3

particular farmers. Near the bottom of Fig. 9.5 it can be seen that the polyhedra BA CW CO and RE CW CO both have three vertices to their polyhedra, and that the polyhedra are connected at that level (the 'I' is at their maximum dimension). This is possible because, although the vertices are not the same (two of them are and the third is different), they are both faces (subsets) of some higher polyhedron in the data set, which connects them. Completely disjoint polyhedra will have no 'I' sign at all. Those that are partially disjoint will have their 'I' sign at some distance to the left (below) their maximum dimension. These polyhedra are said to be 'eccentric'.

A feature noticeable in most of the data sets we have analysed is that the connectivities are high, there are few

Fig. 9.5

Dimension 00000000001111 01234567890123 1	2	Social Class CAP	 RCH	 MID 3	 SML	 MRG
********** I	PA RA BA JO SM MA WH RE CW GR	1	0	0	0	0
********* I	PA RA BA JO MA WH RE CW GR	0	1	0	0	0
******** I	PA RA BA JO SM RE CW GR	1	0	0	0	0
******** I	RA BA SM MA RE CW HO GR	0	0	0	1	0
********* I	PA RA BA SM MA RE CW HO GR	1	0	0	0	0
********** I	PA RA BA JO SM WH RE HO GR CO	1	0	0	0	0
******* I	PA RA BA JO RE CW TO	0	0	0	1	1
******* I	PA RA BA JO RE HO TO	0	0	0	3	2
******** I	PA RA BA JO RE CW HO TO	0	2	0	1	0
********* I	PA RA BA JO WH RE CW HO TO	1	0	0	0	0
********* I	PA RA BA JO WH RE CW TO CL	0	0	0	0	1
******** I	PA RA BA JO RE HO TO CL	1	0	0	0	0
********* I	PA RA BA JO RE CW HO TO CL	1	2	0	0	0
********** I	PA BA JO WH RE CW HO TO CL GR	0	1	0	0	0

(contd. Fig. 9.5)

1	2	3				
******* I	PA RA BA JO RE TO CL	0	1	0	0	0
******* I	PA RA JO SM RE CW GR	1	0	1	0	0
******* I	PA RA BA SM RE HO GR	1	0	0	0	0
****-** I	PA RA SM WH RE HO GR	1	0	0	0	0
******* I	PA RA SM RE CW HO GR	0	0	0	1	0
******* I	RA BA SM RE CW HO GR	0	0	0	1	0
******** I	PA RA BA JO SM RE TO GR	1	0	0	0	0
******* I	PA RA WH RE HO GR CO	1	0	0	0	0
****** I	PA RA BA JO RE CW	0	0	0	2	0
****** I	PA RA BA JO RE HO	0	0	0	1	1
****** I	PA JO WH RE TO CL	1	0	0	0	0
****** I	PA RA BA JO RE GR	1	0	0	0	0
****** I	PA RA BA SM RE GR	0	1	0	0	0
****** I	PA RA JO SM RE GR	1	2	0	1	1
****** I	PA BA JO SM RE GR	0	0	1	0	0
****** I	PA JO SM WH RE GR	0	0	0	1	0

(*contd. Fig. 9.5*)

1	2	3				
****** I	PA RA SM RE CW GR	1	0	0	0	0
****** I	RA BA SM RE CW GR	0	0	0	1	0
****** I	PA JO SM RE CW GR	0	1	0	0	0
****** I	PA RA BA RE HO GR	0	1	0	0	0
****** I	PA RA JO RE HO GR	1	1	0	0	0
****** I	PA RA SM RE HO GR	0	0	0	1	1
****** I	PA BA SM RE HO GR	0	0	0	1	0
****** I	RA BA SM RE HO GR	0	0	0	1	0
****** 1	PA JO SM RE HO GR	0	1	0	0	0
****** I	PA RA WH RE HO GR	1	0	0	0	0
****** I	PA SM RE CW HO GR	0	1	0	0	0
****** I	RA SM RE CW HO GR	0	1	0	0	0
****** I	PA RA SM RE TO GR	1	0	0	0	0
******* I	PA BA SM RE CW GR CO	1	0	0	0	0
****** 1	PA BA RE HO GR CO	1	0	0	0	0
***** I	PA RA BA JO RE	0	0	0	0	1

(*contd. Fig. 9.5*)

1	2	3				
***** I	PA BA JO WH RE	1	0	0	0	1
***** I	PA BA JO RE HO	0	0	0	1	0
***** I	PA JG WH RE TO	1	0	0	0	0
***** I	PA BA JO RE CL	0	0	0	2	0
***** I	PA JO RE TO CL	1	0	0	0	0
***** I	PA WH RE TO CL	1	0	0	0	0
***** I	PA RA SM RE GR	1	0	0	3	1
***** I	PA RA RE CW GR	0	0	0	1	0
***** I	PA BA RE CW GR	0	0	0	0	1
******* I	PA RA MA WH HO TO GR	1	0	0	0	0
***** I	BA JO RE CW CO	0	2	1	2	0
****** I	BA JO SM RE CW CO	0	1	0	0	0
**** I	PA JO WH RE	0	0	0	1	1
**** I	BA JO RE CW	0	0	0	2	0
**** I	PA RA JO HO	0	1	0	0	0
**** I	PA BA JO HO	0	0	0	1	0

(*Contd. Fig. 9.5*)

1	2	3				
****	PA JO RE CL	0	0	0	2	3
I						
****	PA RE TO CL	2	0	0	0	0
I						
****	PA RA SM GR	1	0	0	1	0
I						
****	PA RA RE GR	0	1	0	0	0
I						
****	PA JO RE GR	0	2	0	0	0
I						
****	RA RE CW GR	0	0	0	1	0
I						
****	PA RA HO GR	0	0	0	1	0
I						
****	BA JO RE CO	0	2	0	1	0
I						
****	BA RE CW CO	2	5	3	11	0
I						
*****	BA RE CW HO CO	0	0	0	1	0
I						
***	PA JO RE	0	0	0	1	6
I						
***	PA WH RE	0	0	0	1	0
I						
***	BA JO CW	0	0	1	0	0
I						
***	PA RE CW	0	0	0	0	1
I						
***	PA RA HO	1	0	0	0	0
I						
***	SM RE HO	0	0	0	0	1
I						

(*Contd. Fig. 9.5*)

1	2	3				
*** I	PA JO GR	0	1	0	0	0
*** I	SM RE GR	0	0	1	0	0
*** I	SM HO GR	0	0	0	0	1
*** I	BA RE CO	0	1	0	0	0
°** I	BA CW CO	0	0	0	1	0
*** I	RE CW CO	0	1	0	1	0
** I	PA RA	1	0	0	0	1
** I	PA RE	0	0	0	1	1
** I	BA RE	0	0	0	1	0
** I	BA CW	0	0	0	6	2
** I	RE CW	0	0	0	0	1
** I	RA HO	0	0	0	0	1
** I	RE GR	0	0	0	0	2
** I	RE CO	0	0	0	5	3
* I	BA	0	0	0	1	0
* I	GR	0	0	0	1	0
*	CO	0	0	0	2	0

obvious discontinuities, and the levels of eccentricity are low. All of this means that there are few obstructions to complex relationships affecting directly and indirectly many farmers.

Traffic aggregation

The traffic is shown as the number of farmers of various classes, and total market production etc., for the various combinations of crops. Some combinations involving many crops are rare, whereas other combinations involving some subset of the combinations may be more common. It is, therefore, worthwhile in a number of cases adding the traffic for a high dimension combination to the traffic for a lower dimension combination which is one of its subsets. This offers a certain simplification of the data set—but one with certain properties that must be borne in mind. It might appear *prima facie* that we should generate all the possible combinations of 1, 2, 3 dimensions etc. and search the data for the cumulated traffic values on these. But clearly, this would generate formidable numbers of possibilities, and in fact stretch even the largest computer. The programmes we use cumulate data onto only those lower dimension polyhedra that actually exist already in the data set.

Traffic and backcloth

It ought to be said that Q-analysis was introduced to this study after the data had been collected and organised. The backcloth in most of what follows has been defined as involvement in a crop, either by production, purchasing or selling. Had Q-analysis been introduced from the beginning this would probably not have been accepted as a 'true' backcloth. The backcloth structure is defined by the necessary concomitants of farming, and hence purchasing a crop without producing it, or selling it without producing it (the gains from payment in kind for labour), would appear as traffic on a particular backcloth. Nevertheless, the combinatorial value of the analysis has still vindicated itself.

Here we examine first the matrix in which the 181 farmers are represented by the polyhedra of their own crop combination. (Remember in this instance we are combining purchase, sell, and produce, so that a farmer is involved with a crop by any or all of these).

The crop combinations reveal no major discontinuities. There is an extraordinary number—96 combinations for 181 farmers. They range from three cases of combinations of ten crops to three cases of monocrop involvement. But they are characterised by a high frequency of large combinations. Table 9.3.

TABLE 9.3 : **Frequency of crop combinations**

Number of crops	Number of combinations	% of combinations
10	3	10
⩾9	8	8
⩾8	13	14
⩾7	24	25
⩾6	47	49

Half the combinations are of six or more crops. Most have only one case. There is an observable trend of association between high social class and diversity in crop combination, in contra-distinction to Jodha's observation.[7] Our results may differ from his because we are using production *and* exchange relations to define involvement in crop-economy while Jodha is considering production by itself. Crop combinations in which more than five producers are involved have been underlined in Fig. 9.5. There are only nine. They run the gamut of complexity. One : that of bajra, redgram, cowpea and cotton is grown by 21/181 producers, but its relatively high frequency of occurrence is unique. By and large crop combinations are individualised.

THE IMPORTANCE FOR MARKET INVOLVEMENT OF FREQUENTLY OCCURRING CROP COMBINATIONS

The most commonly occurring crop combination is neither very common, accounting for twelve per cent of our total sample of farmers, nor very complicated even though there is great complexity at the level of the whole data set. It consists of a dry cereal, bajra, two pulses, redgram and cowpea and an agro-industrial crop, cotton. However, amid the diversity of 96 crop combinations, the importance of this one in its contribution to total production, sales and purchases, is nothing short of staggering. In Table 9.4A the production (traffic) figures are given for this polyhedron (backcloth) in total weights and also by percentages. Almost all the bajra produced and commercialised by the capitalist class is produced, sold and purchased in this crop combination. The same is true, though to a lesser degree throughout the system of social classes. Between a quarter to half of the cowpea produced and half to all of the cowpea exchanged on the market by all social classes comes from this crop combination. Redgram and cotton are unimportant for capitalists in this combination but they increase importance in production and even more so in exchange the lower social class. Thus this crucially important crop combination has significance which varies with social class.

It is clear that in this combination foodgrains are an important *commercial* crop. Regional disaggregation, Table 9.4D shows that the combination is mostly restricted to Kovilpatti where crop combinations are simpler than in the other two regions. The class attributes of this combination are also attributes of the region. Lower order combinations of this crop combination, however, do not show such regional specificity.

Cumulating the values for production, sales and purchases of all crop combinations which include all these four key crops with others, or lower order crop combinations including subsets of the four (Table 9.4C) does not increase the importance of any of the four crops by much. Only four further crops become involved (paddy, jowar, small millet, groundnut). The quantitative importance of these four extra crops is slight.

The eight remaining crop combinations used by five or more farm households are shown in Table 9.5. They differ from the first combinations more than they differ between themselves for their class coverage is more restricted and their

TABLE 9.4A : **Crop combination of highest frequency : Production, sales and purchases in kilos of bajra+redgram+cowpea+cotton**

Class of farmer and number of farmers	Capitalist F(2)			Rich F(5)			Middle F(3)			Small F(11)		
	Produce	Sell	Buy	Produce	Sell	Buy	Produce	Sell	Buy	Produce	Sell	Buy
Bajra	6,810	6,000	29	6,884	720	12	915	400	1	4,938	2,210	24
Redgram	230	200	9	1,830	720	3	745	570	3	1,253	1,125	66
Cowpea	485	670	16	415	690	10	116	40	4	430	55	19
Cotton	468	218	42	2,570	240	78	798	785	72	2,032	986	211

TABLE 9.4B : **Crop combination of highest frequency : Production, sales and purchases expressed as a percentage of total production, sales, and purchases for given crops and given classes**

bajra+redgram+cowpea+cotton

Class of farmer % of class	Capitalist F(6%)			Rich F(15%)			Middle F(38%)			Small F(16%)		
	Produce	Sell	Buy	Produce	Sell	Buy	Produce	Sell	Buy	Produce	Sell	Buy
Bajra	77	98	94	52	42	66	45	66	100	29	39	48
Redgram	3	8	5	26	33	2	58	84	100	18	36	19
Cowpea	31	90	100	39	63	45	49	50	40	26	8	51
Cotton	11	7	100	57	24	39	90	89	85	52	43	62

TABLE 9.4 C : **Production of any crop grown in combination with any of bajra, redgram, cowpea, cotton, expressed as a percentage of total production for that crop and any given social class**

	Capitalist F			Rich F			Middle F			Small F		
	Produce	Sell	Buy	Produce	Sell	Buy	Produce	Sell	Buy	Produce	Sell	Buy
Paddy	0.5		3									
Bajra	78	98	94	74	53	83	70	100	100	41	98	58
Jowar				10		9	94		69	16		
Small millet	4			11	100							
Redgram	9	16	1	42	47	6	63	92	100	24	49	20
Cowpea	33	91	100	64	76	86	72	100	100	33	9	57
Groundnut	4	4										
Cotton	19	16	100	72	55	63	100	100	100	68	52	75

TABLE 9.4 D : **Cumulative percentage of farmers in each region involved in bajra+redgram+cowpea+eotton in given subsets (even in combination with unnamed crops)**

Number of farmers region	(52) Hyderabad	(60) Kovilpatti	(69) Anantapur
Bajra	53	77	29
Bajra redgram	53	58	29
Bajra cowpea	25	66	13
Bajra cotton	—	—	—
Redgram	26	55	26
Redgram cotton	—	70	6
Cowpea cotton	—	—	—
Bajra redgram cowpea	—	—	—
Bajra cowpea cotton	—	48	1
Redgram cowpea cotton	—	50	1
Bajra redgram cotton	—	53	4
Bajra redgram cowpea cotton	—	47	1

TABLE 9.5 : **Crop combinations used by more than four farmers : Percentage of total for given class and given crop**

	Capitalist F			Rich F			Middle F			Small F (3)			Marginal F (2)		
	Produce	Sell	Buy	Produce	Sell	Buy	Produce	Sell	Buy	Produce	Sell	Buy	Produce	Sell	Buy
1		2			3			4			5			6	
Paddy												14			9
Ragi										17			30		
Bajra										4			27		
Jowar										18		38	20		32
Redgram										2	3	13	0		12
Horsegram										16	37		14		
Tomato										18	17	100	34	39	100
	Capitalist F (1)			Rich F (2)			Middle F			Small F (1)			Marginal F (1)		
Paddy	2	1		11		26				3			10		
Ragi	2			11						1			4		30
Jowar	4	1	4	7	12					10					6
Small millet	9			9						10			39		
Redgram	9	19		4						2			9	17	
Groundnut	3	3		8	7					2			5	24	40

TABLE : 9.5 (*Contd.*)

	Capitalist F (1)			Rich F			Middle F			Small F (3)			Marginal F (1)		
Paddy	5	1								14		1	2		
Ragi	3									9		8	1		
Small millet	9									18			13		
Redgram	3	4								9	8		2	2	
Groundnut	5	4								11	12	19	11	11	
	Capitalist F			**Rich F (2)**			**Middle F (1)**			**Small F (2)**			**Marginal F**		
Bajra				12		16	25	33		9	5	10			
Jowar				10		9	94			16		1			
Redgram				11	1	4	5	7	62	3	1	1			
Cowpea				25	14	18	23	57	60	4		4			
Cotton				9		13	16	10	12	9		9			
	Capitalist F			**Rich F**			**Middle F**			**Small F (2)**			**Marginal F (6)**		
Paddy												20			22
Jowar										2		5	3		4
Redgram												11			15
Cluster bean										61	60	62	59	35	66

TABLE : 9.5 (*Contd.*)

1	2	3	4	5			6		
	Capitalist F	Rich F	Middle F	Small F (1)			Marginal F (6)		
Paddy				5		12			30
Jowar						5	12		11
Redgram						7			29
	Capitalist F	Rich F	Middle F	Small F (6)			Marginal F (2)		
Bajra				16	8	24	44		100
Cowpea				10	28	8	13		66
	Capitalist F	Rich F	Middle F	Small F (5)			Marginal F (3)		
Redgram				11	16	10	6	9	2
Cotton				22	30	14	100	100	100

quantitative importance is reduced. Since small farmers are the largest class category it is not surprising that they and marginal farmers are the classes in which commonly occurring combinations are found. What may be surprising is that these classes appear to dominate complex common combinations as well as simple ones. All but two of these combinations, however simple, include a market-oriented crop and at least one pulse and one cereal.

Table 9.9 has three principal features :

(i) The low frequency of occurrence of even the more common crop combinations which points once more to the specificity of combinations.

(ii) In almost every detail these crop combinations are most important in terms of their contribution to production, sales and/or purchase of the crops concerned than they are in terms of their frequency of occurrence. The implication here must be that most of the 87 remaining combinations do not contribute much to the system of exchange.

(iii) Crop combinations may nonetheless involve striking emphases on a subset of one or a few crops (ragi, jowar, horsegram, tomato in the 7-combination ; jowar and cowpea in the 5-combination ; paddy in the 4 and 3-combinations and bajra and cotton in the 2-combinations.)

Lastly, Table 9.6 shows the characteristics of the most highly complex crop patterns and those of the most simple ones. In all but one instance these are case studies of individual farms. The three 10-combinations are all from the labour-employing classes. Production patterns are characterised by heavy concentration on one or two crops for the market and by production of petty quantities of large numbers of crops which are not sold. Dependence upon the market for purchases is slight in absolute terms. These single individuals may have considerable power on the market. One capitalist farmer in Anantapur sells 27 per cent of all cotton sold by capitalists, 14 per cent of all groundnut and 29 per cent of all

TABLE : 9.6 **Most complex and simplest crop combinations**
Percentage of total for given class and given crop

	Capitalist (1) Anantapur			Capitalist (1) Anantapur			Rich peasant (1) Hyderabad		
	Produce	Sell	Buy	Produce	Sell	Buy	Produce	Sell	Buy
Paddy			17	4	3		1		4
Ragi	4		20	14		10			
Bajra	1			3			1		
Jowar	1			5			3		27
Small millet	2			15					
Maize	62			17	29		52		
Wheat	17			7	4				11
Redgram	6	4							
Cowpea	48							5	
Horsegram				4				3	
Tomato							20	16	
Cluster bean							12		
Groundnut	5	4		12	14		12		
Cotton				23	27				

	Small peasant (1) Anantapur			Small peasant (1) Anantapur			Small peasant (2) Kovilpatti		
Bajra				1	4				
Groundnut	10	12							
Cotton							3	3	2

TABLE 9.7 : **Aggregate production, sales and purchases by social class**

	Capitalist F (34)			Rich F (33)			Middle F (8)			Small F (70)			Marginal F (36)		
	Pro-duce	Sell	Buy	Pro-duce	Sell	Buy	Pro-duce	Sell	Buy	Pro-duce	Sell	Buy	Pro-duce	Sell	Buy
Paddy	1,29,667	14,660	2,393	22,416	700	5,688	700	0	220	11,685	790	11,770	4,128	1,000	11,826
Ragi	15,080	600	1,000	6,780	400	210	192	0	200	4,257	0	1,625	2,680	0	300
Bajra	8,891	6,100	31	13,179	1,720	18	2,045	600	1	17,048	5,625	50	1,512	0	5
Jowar	4,554	0	360	18,468	1,600	758	3,190	0	160	10,950	270	1,597	2,756	0	1,626
Small Millet	11,055	0	0	5,642	500	0	745	0	0	5,276	0	0	1,165	0	0
Maize	620	0	0	85	0	0	0	0	0	120	0	0	0	0	0
Wheat	2,241	340	35	108	100	0	0	0	0	103	100	25	50	0	15
Redgram	8,689	2,610	171	7,056	2,144	136	1,280	675	3	6,798	3,139	344	1,979	1,170	242
Cowpea	1,406	741	16	1,061	1,098	22	236	80	10	1,634	625	37	331	175	3
Horsegram	3,965	1,650	0	1,255	575	0	0	0	0	2,844	1,070	0	620	130	0
Tomato	5,730	5,072	14	5,791	6,211	14	0	0	0	2,941	2,837	3	1,992	1,982	4
Clusterbean	4,030	3,943	17	804	389	1	0	0	0	1,129	1,139	8	1,222	730	3
Groundnut	61,262	43,997	400	22,770	16,455	420	2,520	1,200	0	16,333	9,700	939	3,080	2,146	200
Cotton	4,331	3,576	42	4,525	990	200	887	845	82	3,889	2,308	341	165	150	9

Note : All crops are not produced by all producers.

wheat. One rich peasant in Hyderabad sells 16 per cent of all tomatoes sold by rich peasants. By contrast, all monocropping is done by small peasants. From Table 9.6 we can also see that monocropping is one of two types. Firstly, there are cases in Anantapur and Kovilpatti of specialisation in one agro-industrial crop, all of which is sold, presumably for cash for food. Secondly, there is a case in Kovilpatti of production, and purchase, of a single cereal. The purchase is presumably paid for by income from wage labouring.

The first type is more important quantitatively than is the second. This suggests very small farmers may be pushed out of a subsistence strategy into more intensive commercial landuse.

MARKET INVOLVEMENT AGGREGATED : MARKETED SURPLUS AND MARKET DEPENDENCE

Simple non-combinatorial aggregation shows for each crop the marketed and net marketable surplus by class of producer (Table 9.7). Table 9.8 shows simple totals by crops alone.

TABLE 9.8 : **Total production, sales and purchases (kgs)**

	Production	Sales Marketed Surplus (% production)	Purchases	Market dependence (% prod.)	 (% sales)
Paddy-rice	1,68,596	17,150 (10)	32,527	(19)	(190)
Ragi	28,989	1,000 (3)	3,335	(11)	(333)
Bajra	42,675	14,045 (33)	105	(0)	(0)
Jowar	39,918	1,870 (5)	4,501	(11)	(240)
Small millet	23,883	500 (2)	—	(—)	(—)
Maize	825	— (0)	—	(—)	(—)
Wheat	2,502	540 (22)	75	(3)	(13)
Redgram	25,802	9,738 (38)	896	(3)	(9)
Cowpea	4,668	2,719 (58)	88	(2)	(3)
Horsegram	8,684	3,485 (40)	—		
Tomato	16,454	16,102 (98)	35	(0)	(0)
Clusterbean	7,185	6,201 (87)	29	(0)	(0)
Groundnut	1,05,965	73,498 (70)	1,959	(2)	(3)
Cotton	13,797	7,867 (57)	674	(4)	(9)

Predictably enough, the marketed surplus of the market-oriented crops is highest, varying from 57 to 98 per cent (Table 9.8). Small and marginal farmers sell a greater proportion than middle and rich farmers, a proportion on a par with that sold by capitalists. We have no hard data to support our speculative conclusion for this trend—but whereas the very high marketed surplus of the capitalist may be the result of a price responsive commercial specialisation, that of the small farmers may represent the compulsion to commercialise, as noted above.

Pulses also have high market surplus, varying from 38 to 58 per cent. The surplus appears perverse, being higher among the labouring classes than among the exploiting ones. The marketed surplus of cereals shows the greatest variety. For bajra it is high (33%), capitalists selling a proportion twice as large as marginal peasants. For wheat it is lower, 22 per cent though the class aggregation shows that production by rich and middle peasants is totally for the market. For paddy anc rice the marketing surplus is low—10 per cent—and, lik pulses, with marginal farmers selling the greatest proportio of production. For the other cereals the marketed surplus i negligible.

Poor producers are often characterised as self-sufficier in basic foodgrains, with rich producers specialising in accor with the theory of comparative advantage. The latter ma grow and sell agro-industrial crops in such a way that they ar dependent upon purchases from the market for their provisior ing. Our evidence convincingly refutes this characterisatio (Tables 9.9 and 9.10A, B, and C). In every case with th exception of tomatoes and clusterbeans, but including th superior grains and the other market-oriented crops th quantities bought by the small and marginal peasantry excee those purchased by the rich peasantry and by capitalists. A Hill[8] remarks "the richest cultivators are the only re: subsistence farmers".

TABLE 9.9 : Marketed and marketable surplus by class

	Capitalists		Rich F		Middle F		Small F		Marginal F	
	Marketed surplus	Marketable surplus	(1)	(2)	(1)	(2)	(1)	(2)	(1)	(2)
Paddy-rice	11	9	3	—22	0	—31	7	—94	24	—262
Ragi	4	—3	17	1	0	—104	0	—38	0	—11
Bajra	69	68	13	13	29	29	32	32	—	0
Jowar	0	—8	9	5	0	—5	2	—12	0	59
Small millet	0	0	9	9	0	0	0	0	0	0
Maize	0	0	0	0	0	0	0	0	0	0
Wheat	15	14	92	0	0	0	97	72	0	—30
Redgram	30	28	30	28	52	51	44	41	59	47
Cowpea	52	51	103	101	33	29	38	36	53	53
Horsegram	42	42	45	45	0	0	38	38	31	31
Tomato	88	88	107	107	0	0	96	96	99	99
Clusterbean	98	98	48	48	0	0	100	100	60	60
Groundnut	72	71	72	70	48	48	59	54	70	63
Cotton	82	81	22	17	95	86	59	50	91	85

(1) Marketed surplus—sales as per cent production.
(2) Marketable surplus—sales net of purchases as per cent production.

TABLE 9.10A : **Percentage share of classes in production**

	Capitalists	Rich F	Middle F	Small F	Marginal F
Paddy-rice	77	13	4	7	2
Ragi	52	23	1	15	9
Bajra	21	31	5	40	3
Jowar	11	46	8	27	7
Small millet	46	24	3	22	5
Maize	75	10	—	15	—
Wheat	90	4	—	4	2
Redgram	34	27	5	26	7
Cowpea	30	23	5	35	7
Horsegram	46	14	—	33	7
Tomato	35	35	—	18	12
Clusterbean	56	11	—	16	17
Groundnut	58	21	2	15	3
Cotton	22	33	6	28	1
% of producers	19	18	4	39	20

TABLE 9.10B : **Percentage share of classes in sales**

	Capitalists	Rich F	Middle F	Small F	Marginal F
Paddy-rice	85	4	—	5	6
Ragi	60	40	—	—	—
Bajra	43	12	4	40	—
Jowar	—	86	—	14	—
Small millet	—	100	—	—	—
Maize	—	—	—	—	—
Wheat	63	18	—	18	—
Redgram	27	22	7	32	12
Cowpea	27	40	3	23	6
Horsegram	47	16	—	31	5
Tomato	31	38	—	18	13
Clusterbean	64	6	—	18	12
Groundnut	60	22	2	13	3
Cotton	45	12	11	29	2

Percentage of sellers

TABLE 9.10C : **Pe:·centage share of classes in purchases**

	Capitalists	Rich F	Middle F	Small F	Margioal F
Paddy-rice	7	17	6	36	36
Ragi	30	6	6	49	9
Bajra	29	17	1	48	5
Jowar	8	17	—	6	36
Small millet	—	—	—	—	—
Maize	—	—	—	—	—
Wheat	47	—	—	33	20
Redgram	19	15	0	38	27
Cowpea	18	25	11	42	4
Horsegram	—	—	—	—	—
Tomato	40	40	—	9	11
Clusterbean	59	3	—	28	10
Groundnut	20	21	—	47	10
Cotton	6	30	12	51	1

Percentage of purchases

That this is the case emphasises the political importance of the behavioural efficiency of markets. That the labour employing classes are less dependent on commodity markets may indicate the consequence of interlocked commodity and labour markets. Commodities are withheld from the market in order to pay wage-labour in kind.

Table 9.11 disaggregates total production, sales and purchases for each crop by region. The cropping pattern of Kovilpatti villages is the simplest (6/14 crops) while those of Hyderabad and Anantapur are more complex (12/14 and 14/16 respectively). There is no regional difference in the high marketed surplus or in the social distribution of surplus for the market-oriented crops, except for cotton which is the preserve of the capitalist class in Anantapur. Pulse production tends to be for subsistence in Anantapur while it is highly commercialised in Kovilpatti. The social trends, however, show no strong regional differences. The dependence of small and marginal peasants on *purchased* foodgrains shows strong regional specificity : rice, wheat and jowar in Hyderabad, bajra in Kovilpatti and ragi in Anantapur.

TABLE 9.11 : **Aggregate production, sales and purchases, by social class and region**

(*A*) *Anantapur*

	Capitalist F (10)			Rich F (6)			Middle F (0)			Small F (17)			Marginal F (19)		
	Produce	Sell	Buy	Produce	Sell	Buy	Produce	Sell	Buy	Produce	Sell	Buy	Produce	Sell	Buy
Paddy	66,162	10,860	1,033	2,405	0	2,188	0	0	0	8,020	500	9,725	2,680	1,000	11,776
Ragi	1,180	600	0	2,890	0	0	0	0	0	3,139	0	0	2,305	0	0
Bajra	709	100	2	1,270	0	0	0	0	0	1,854	0	5	832	0	0
Jowar	2,045	0	341	3,305	0	688	0	0	0	5,325	0	1,523	2,756	0	1,526
Small millet	0	0	0	0	0	0	0	0	0	0	0	0	0	0	0
Maize	0	0	0	0	0	0	0	0	0	0	0	0	0	0	0
Wheat	435	40	35	56	0	0	0	0	0	53	0	25	50	0	15
Redgram	66	10	162	161	39	103	0	0	0	374	300	241	54	0	237
Cowpea	140	71	0	274	230	0	0	0	0	568	568	0	244	150	0
Horsegram	145	50	0	450	350	0	0	0	0	834	660	0	92	40	0
Tomato	5,516	5,072	14	5,791	6,211	14	0	0	0	2,941	2.837	8	1,992	1,982	4
Clusterbean	4,030	3,943	17	804	389	1	0	0	0	1,129	1,139	8	1,222	730	3
Groundnut	0	0	0	50	0	0	0	0	0	0	0	0	0	0	0
Cotton	0	0	0	0	0	0	0	0	0	0	0	0	0	0	0

TABLE 9.11 : (*Contd.*)

(*B*) *Kovilpatti*

	Capitalist F (2)			Rich F (12)			Middle F (5)			Small F (35)			Marginal F (6)		
	Produce	Sell	Buy	Produce	Sell	Buy	Produce	Sell	Buy	Produce	Sell	Buy	Produce	Sell	Buy
Paddy	0	0	0	0	0	0	0	0	0	0	0	0	0	0	0
Ragi	0	0	0	0	0	0	0	0	0	0	0	0	0	0	0
Bajra	6,810	6,000	29	11,654	1,720	18	1,985	600	1	14,778	5,625	45	660	0	5
Jowar	0	0	0	3,350	200	70	3,150	0	110	4,500	270	24	0	0	0
Small millet	0	0	0	600	500	0	0	0	0	0	0	0	0	0	0
Maize	0	0	0	0	0	0	0	0	0	0	0	0	0	0	0
Wheat	0	0	0	0	0	0	0	0	0	0	0	0	0	0	0
Redgram	2,230	200	9	3,782	1,205	33	810	620	3	2,668	2,194	103	145	100	5
Cowpea	435	670	16	680	850	22	216	80	10	948	235	37	42	0	3
Horsegram	0	0	0	0	0	0	0	0	0	0	0	0	0	0	0
Tomato	0	0	0	0	0	0	0	0	0	0	0	0	0	0	0
Clusterbean	0	0	0	0	0	0	0	0	0	0	0	0	0	0	0
Groundnut	0	0	0	0	0	0	0	0	0	0	0	0	0	0	0
Cotton	468	218	42	4,525	990	200	887	845	82	3,889	2,306	341	165	150	9

(C) Hyderabad

	Capitalist F (22)			Rich F (15)			Middle F (3)			Small F (18)			Marginal F (11)		
	Produce	Sell	Buy	Produce	Sell	Buy	Produce	Sell	Buy	Produce	Sell	Buy	Produce	Sell	Buy
Paddy	63,505	2,800	1,360	20,011	700	3,500	700	0	220	3,665	290	2,045	1,448	0	50
Ragi	13,900	0	1,000	3,890	190	219	192	0	200	1,064	0	1,625	375	0	300
Bajra	1,372	0	0	255	0	0	60	0	0	416	0	0	20	0	0
Jowar	2,509	0	16	11,813	1,400	0	40	0	50	1,125	0	50	0	0	100
Small millet	11,055	0	0	5,042	0	0	745	0	0	5,276	0	0	1,165	0	0
Maize	620	0	0	85	0	0	0	0	0	120	0	0	0	0	0
Wheat	1,860	300	0	52	100	0	0	0	0	50	100	0	0	0	0
Redgram	8,393	2,400	0	3,110	900	0	470	55	0	3,756	645	0	1,780	1,070	0
Cowpea	831	0	0	107	18	0	20	0	0	118	36	0	45	25	0
Horsegram	3,820	1,600	0	805	225	0	0	0	0	2,010	380	0	528	150	0
Tomato	214	0	0	0	0	0	0	0	0	0	0	0	0	0	0
Clusterbean	0	0	0	0	0	0	0	0	0	0	0	0	0	0	0
Groundnut	61,262	43,997	400	22,720	16,455	420	2,520	1,200	0	16,333	9,700	939	3,080	2,146	200
Cotton	3,863	3,358	0	0	0	0	0	0	0	0	0	0	0	0	0

CONCLUSIONS

The application of Q-analytical methods to the problem of complexity in crop combinations in the dryland agricultural economy has allowed both a high compression of data and the retention of complexity.

In this case all existing crop combinations have been computed. Their frequencies of occurrence, their contributions to production, sale and purchase (market involvement) individually and cumulatively, by class and by region, have been calculated.

Our substantive conclusions will be summarised under four headings : those about crops, those about class, those about region and implications for policy.

Crops and crop combinations

(i) of the 14 crops we have considered paddy-rice and redgram are the crops most highly connected and may be considered the core crops of the micro-system of production and exchange. Market oriented crops are least well connected, and involved with other crops in most regionally specific ways ;

(ii) the patterns of crop combination in dryland agriculture are highly diverse and individualised. 181 farmers had 96 crop combinations ;

(iii) these patterns are also highly complex. Forty nine per cent of combinations were of six or more crops ;

(iv) only 9 of the 96 combinations involved the participation of five or more households. One combination (bajra, redgram, cowpea and cotton) characterised 12 per cent of our sample. This combination dominates the market involvement of the respective crops, but in different ways across social classes ;

(v) characteristic of the eight other 'commonly' occurring combinations is a concentration on one crop (often a commercialised crop) and petty production for subsistence of the other crops ;

(vi) this also characterises the most complicated combinations of ten crops ;

(vii) monocrop cultivation is of two sorts—firstly complete specialisation for the market of an agro-industrial crop ; secondly complete specialisation on a dry cereal for subsistence, together with its purchase ;

(viii) the marketed surplus of dryland crops is highest for market-oriented (agro-industrial) crops and for pulses. It is lowest and variable for cereals. Most farmers sell only small amounts ; and

(ix) market dependence is non-existent for maize, small millet and horsegram. It is negligible and/or for seed for market-oriented crops and for bajra and ragi. It exceeds the marketed surplus in the cases of paddy-rice and jowar.

Class characteristics of market involvement and crop combinations

(i) the labour-employing classes have more complex crop combinations than do the labour providing classes ;

(ii) capitalists and rich farmers are most frequently involved in the market for most crops ;

(iii) in the case of 'superior' cereals—paddy-rice and wheat—capitalists produce for the market while the labouring peasantry are dependent on the market for purchases ;

(iv) the labouring peasantry is significantly dependent on the market for purchase for consumption of all cereals ;

(v) single individuals (especially capitalists and rich peasants) can be major suppliers to markets. We cannot pursue the implications rigorously here because of inadequacies of data, but it is likely that village commodity markets are dominated by single or very few suppliers from the labour-employing classes ;

(vi) though these classes generally dominate production, they dominate market supplies even more. They are powerful controllers of grain. They are disproportionately less dependent upon the market for purchases for consumption than either their numbers or their contribution to production would suggest. They are more self-sufficient in foodgrains ;

(vii) however, this domination is less marked with respect to market-oriented crops, where the labouring peasantry are significant producers and suppliers to the market ;

(viii) in the case of pulses and paddy-rice marketed surplus (the proportion of production which is sold) the labouring classes sell a higher proportion than do the labour-exploiting classes. For the former, these crops take on the characteristics of agro-industrial crops rather than food. The latter may well dispose of their crop on the labour market—as kind payment to wage labour—rather than on the commodity market ; and

(ix) of 12 out of the 14 crops, and including the superior grains, the absolute quantities purchased by the labouring peasantry exceed those purchased by the labour-exploiters.

Regional characteristics of market involvement and crop combinations

(i) regional patterns of crop combination differ in complexity. Those of Kovilpatti are relatively simple. Those of Hyderabad and Anantapur are highly complex ;

(ii) each region is characterised by core crops and combinations connected with which are most other crops. Paddy and redgram are core crops in Hyderabad—bajra, redgram, cowpea and cotton are core crops in Kovilpatti—paddy, ragi, groundnut and horsegram are core crops in Anantapur.

(iii) the 4-combinations (bajra, redgram, cowpea and cotton) totally dominated production, sale and purchase, of these crops in Kovilpatti, both in isolation and cumulatively. Lower order cumulated crop combinations of these four crops do not show regional specificity ; and

(iv) there are strong regional differences in whether specific rainfed cereals and pulses are produced for subsistence (and purchased) (ragi and bajra in Anantapur and bajra in Hyderabad), or whether they are produced for sale (bajra in Kovilpatti).

REFERENCES

1. International Crop Research Institute for the Semi-Arid Tropics, (Hyderabad).
2. All India Co-ordinated Research Project on Dryland Agriculture.
3. Atkin R.H. (1974). *Mathematical Structure in Human Affairs*, Heinemann, London.
4. Johnson J.H. (Guest Editor) (1981). "Special Issue on Q-Analysis", *Environment and Planning*, Vol. 8, No. 4, Pion (London).
5. G.P. Chapman, "Q-Analysis" in R.J. Bennett and N. Wrigley (Eds.). *British Quantitative Geography*, Routledge, Kegan, and Paul, (London).
6. P.R. Gould, J.H. Johnson, and G.P. Chapman (1983). *Television : The World of Structure*, Pion (London).
7. N.S. Jodha (1980) "Some dimensions of traditional farming systems in semi-arid tropical India, ICRISAT, pp. 11-24.
8 P. Hill (1980). *Dry Grain Farming Families : Hausaland (Nigerial)) and Karnataka (India) Compared*, Cambridge University Press, (Cambridge).

10

CHRISTALLER REVISITED OR DESCRIPTION VERSUS PRESCRIPTION

With Special Reference to Service Provision and Regional Planning in India*

B.H. FARMER

It is a great pleasure and privilege to be asked to contribute to the volume that is being prepared in honour of Professor V.L.S. Prakasa Rao, for whom I have a great affection as a person and a great respect as a geographer of eminence and distinction—indeed, of world stature.

This essay will be in the nature of a personal statement or testament, beginning with a narrative. It is now nearly half a century since I became interested in functional difference between towns and villages and, in some sense, in the existence and characteristics of the urban system. This was in connection with a dissertation that I wrote as an undergraduate in 1936, when I was a candidate for Part I of the Geographical Tripos, then the second-year examination for Cambridge geographers. I called my dissertation 'Malmesbury: A Study in Urban Geography' (or, in a later version presented unsuccessfully! — for an Essay Prize given by the Royal Geographical Society. 'Malmesbury : The Decline of a Country Town'). This juvenile piece of work concerned my birth-

*At the time of writing this paper, the author was not aware that Prof. Prakasa Rao has also contributed a paper with an identical title,

place, an historic small town in north-west Wiltshire, and was at pains to demonstrate that what distinguished Malmesbury from the surrounding villages was a question of function; for Malmesbury acted, in a humble way, as the centre for the life of these villages: at such pains, indeed, that the examiner, the late W.V. Lewis, underlined its later appearances and pencilled 'function, function, function!' in the margin. I chose my alternative title because it was evident from my researches that a number of these functions, notably centrality for cattle marketing, shopping and administration, were being eroded because of the competition, strengthened by the railway and even more by developing road transport, of larger towns, particularly Swindon, only fifteen miles away and then, with some 60,000 people, over twenty times the size. I predicted further decline, and was able to demonstrate some of the visible signs of this when I had the pleasure of taking the six participants in the Second Indo-British Seminar (who included one of the principal inspirers of the Seminar, Prakasa Rao) to Malmesbury in 1975. The town had, however, grown in population and had acquired a number of "non-basic" industries. Perhaps I shall be able to re-survey my native Borough fifty years on from my original effort, in 1984-85.

I hasten to stress that I claim no originality whatsoever for my very earliest piece of research (or for much that came later for that matter). I then knew nothing of Walter Christaller, whose work *Die zentralen Orte in Süddeutschland* (1933), had appeared a year or two before, but, of course, in the original German. But I had read, under the stimulus of H.C. Darby, Lucien Febvre's *A Geographical Introduction to History* (1925), prescribed as a "possibilist" antidote to environmental determinism, and knew that Febvre emphasised that in the study of towns "function is of primary importance" (p. 339). I was also aware and seem to remember, that I modelled much of my work on the late R.E. Dickinson's studies of East Anglia (Dickinson, 1932 and 1934). And, as a West Country-man, I had read, and marked Thomas Hardy's description, now nearly a century old, of the functions of a country town:

> Casterbridge is in many respects but the pole, focus or nerve-knot of the surrounding country life, differing from... manufacturing towns which are as foreign bodies set down like boulders in a plain (*The Mayor of Casterbridge*, 1886).

For Casterbridge lived, as Malmesbury lived in the 1930s, by agriculture "at one remove further from the fountainhead than the adjoining villages, no more". And I saw the deadly rival, Swindon, as one of Hardy's "boulders in a plain" (a matter to which I shall return).

I should also perhaps stress, before I leave the 1930s and my youthful endeavours, that my work was, like the other English works I have cited, purely descriptive, though that did not prevent it, or them from being analytical, or from postulating causes. But it was in no way prescriptive; decline had set in, and would continue, and no suggestion was made that anything could or should be done about it. The brave new world of post-war planning was ten years away.

During those ten years, a wartime posting took me to India; without it I suppose that I should never have met Prakasa Rao, or been influenced by him, or devoted most of my subsequent research and writing to South Asia (nearly all of it, however, on agrarian rather than urban topics). I did return briefly to my native Wiltshire, in mind if not in body, to read H.E. Bracey's *Social Provision in Rural Wiltshire* (1952, p. 132), which did briefly draw attention to Christaller's *Die zentralen Orte in Süddeutschland* and claim that several other writers had advanced similar theoretical frameworks. But it was not until after the appearance of Carlisle W. Baskin's English translation (*Central Places in Southern Germany*) in 1966, that Christaller's ideas were borne to shores that had never previously encountered them: this clearly applies both to India and to Britain. By this time I was deep in agrarian work, though not unaware of problems connected with market centres in, for example, colonization schemes.

Thus I wrote :

> One might expect the initial planning of major colonization schemes to involve the co-ordinated layout not only of transport networks but also of a hierarchy of marketing centres, in order to ensure a high degree of efficiency in the system and hence a maximum stimulus to the development of a market economy and to productivity generally (Farmer, 1974, p. 200).

I remember boring the First Indo-British Seminar held in Delhi in 1972 under the inspiration and administration of Prakasa Rao, with a similar theme, which perhaps owes something to Christaller but more, I think, to Dickinson, Bracey, A.E. Smailes (for example, Smailes, 1944, 1947) and perhaps even to my own juvenilia. With a decade of hindsight, my remarks to the seminar appear not a little naive, as the sequel will show. If I am wiser now about these matters, it must be largely because of the stimulus to critical thought supplied by the successive Indo-British Seminars, and especially by contacts there and elsewhere with Prakasa Rao, K.V. Sundaram and Sudhir Wanmali.

It was Sudhir Wanmali who was responsible for my first really intimate contact with Christaller. This may surprise him, for when he came to Cambridge in 1975 to work as a research student under my supervision, already with a distinguished early career behind him, I extracted from him a promise not to mention the name of Christaller for a year, on pain of buying me a pint of beer on each occasion when the name of the German geographer passed his lips. This was because I was at that time disturbed by what seemed to me the uncritical (or insufficiently critical) application of Christaller's model, descriptively and even more prescriptively, to situations in Third World countries to which it was dangerously irrelevant, or at least insufficiently relevant. Wanmali will, I hope, not mind my saying that I thought that this stricture applied, *inter multa alia*, to some of the works with which he had been associated: for example, *Planning Rural Growth*

Centres for Integrated Area Development : A Study in Miryalguda Taluka (Sen, Lalit K., Wanmali, S. *et al.*, 1971). But it is only fair to him to add that Wanmali was critical of Christaller in some of his publications (e.g. Wanmali, 1973, on behavioural grounds); and that he tells me that he always "opposed the hexagonal geometry and...the *k* values of Christaller's central place theory" : and, while believing "that some of the concepts of central place theory are "application oriented" finds "its geometry and *k* values of little significance in applied analysis". It is, perhaps, fair to me to add that I was, as he well knew, not very serious in forbidding mention of Christaller : for I believe very strongly in interaction between a supervisor and a research student on a basis of intellectual equality, as distinct from a 'top-down' relationship. In academic enquiry we are all students. If we are not, we have no business to be in universities at all.

(Perhaps I should say, in parenthesis, that Christaller is not the only theoretician to be taken out of context and applied all too widely. That early nineteenth-century German *junker*, J.H. von Thünen, whose writing on agricultural location was disinterred by geographers in the 1960s, would be astonished if he were resurrected and could read some of his latter-day applications (Von Thünen, J.H., 1826; Chisholm, M., 1979). Again, geographers have found similarities between the well-known models of Gunnar Myrdal and A.O. Hirschman in their applicability to regional inequalities of income, apparently oblivious to the indisputable fact that, whereas Myrdal was writing as a socialist who deplored inequalities (he was principally concerned with these internationally), Hirschman was writing as a neo-classical economist, a free-enterprise man, who positively advocated regional inequality as a strategy in the early stages of economic development (Myrdal, G.H., 1957; Hirschman, A.O., 1958; Keeble, D.E:. 1967, especially pp. 257-63). Geographers ignore *political* economy at their peril.

Now, at about the time that Wanmali and I began to interact in Cambridge, Harold Brookfield published his seminal book, *Interdependent Development* (1975). I was particularly interested to read what he had to say (pp. 105-10) on growth poles and central places, and, in ensuing pages (pp. 110-23) on theories and ideas on the diffusion of innovation and on 'modernisation'; on the theoretical confusion and conflation that had crept in over time (not very much time, either); on E.A.J. Johnson's well-known and sadly influential conclusion that rural development in India as elsewhere demanded the bringing into existence of a complete spatial system of urban centres arranged in a hierarchy from agro-urban towns through several intermediate types to the metropolis, which clearly reflects his reading of Christaller, though Lösch has also in evidence in his book (NCAER, 1965; Johnson, E.A.J., 1970; Lösch, A., 1954); on T. Hermansen's confessedly critical attempts to relate theories on growth poles, central places, and the diffusion of innovations (Hermansen, T., 1972 and 1975): but above all on his (Brookfield's) rather cautious conclusion that this sort of conflation of theories 'requires a dynamization of central place theory that it does not possess' (p. 109), which struck a responsive chord in my mind.

For scepticism mounted in my thinking, fuelled partly by continued interaction with Wanmali, and by a feeling of unease about a number of efforts to conflate and to dynamise these theoretical strands, in an Indian context and in works that will be familiar to most readers of this essay, so that an extended review is unnecessary (see, for example, Misra, R.P., Prakasa Rao, V.L.S. and Sundaram, K.V.; 1970; Misra, R.P., 1971; Misra, R.P., Sundaram, K.V. and Prakasa Rao, V.L.S., 1974; also NCAER, 1965 and, in a more general context. UN, 1976, p. 25 and Mabogunje Akin L., 1980, pp. 201-6).

But what finally sent me back to Christaller himself and justifies the title of this essay was an invitation early in 1980 to address a Study Seminar, organised by Michael Lipton at the Institute of Development Studies, Brighton, England. The theme of the Seminar was 'Planners, Peasants and Poverty: Rural-Urban and Farm-Nonfarm Relations: the Development Process', a theme clearly related to Lipton's important and influential work on urban bias in relation to rural poverty (Lipton, M., 1977) but which also carried my mind to Prakasa Rao and the Indo-British Seminars on rural-urban relations. I was specifically to speak on ' "Rural" and "Urban" human settlement types and changes'. This clearly involved questions of definition and the controversy on continuum *versus* dichotomy; but soon brought me to Christaller and to careful reading of this book—in English translation I hasten to add (Christaller, 1966).

So I come at last to 'Christaller revisited'. I commend such a contact to all who only know of the sage from summaries in secondary works which, admirable though many of them may be, are no substitute for the sage himself, even though some of his power and perhaps some of his clarity are reduced in translation. The points that particularly struck home to me, in the context of the present essay and of efforts to dynamise his theory and conflate it with growth pole and modernisation theories, and in barest outline, are these (I include some familiar, perhaps too familiar points, to aid clarity):

(1) Christaller (1893-1969) was brought up intellectually in the Germany of the Weimar Republic, after World War I and before Hitler ; and was steeped in the doctrines of the liberal classical economist (his teachers included Robert Gradmann of the University of Erlangen and, it would appear, Alfred Weber). His basic assumptions were of a free market and of perfect competition. This is assuredly an important point if one tries to apply Christaller's ideas through state intervention, as has been so widely advocated in India and elsewhere.

(2) He distinguished (and to this extent was a dichotomist) 'central places' which serve an area, and 'dispersed settlements' which only serve themselves; but, he said, these are not respectively equivalent to town and village because not all towns are central places (compare Hardy's 'boulders'), whereas some villages are; though it is, of course, true that most towns are central places, or become them with the passage of time (a matter to which I shall return)—to this extent Hardy's figure of speech is elegant but inapt.

(3) Christaller was at pains to claim that his model was deductive and stress that it was static; though he clearly had an eye on South Germany, to which in his book, he

later applied it. In fact, I, and others whom I have consulted, conclude that the model is *inductive*, so closely does it follow from an empirical study of Southern Germany, with all its particularities of history and geography. As J.N. Keynes, the father of the great economist, J.M. Keynes, said

> The so-called deductive method in its complete form is thus seen not to be an exclusively deductive method. It. . . is still aided and controlled by induction (Keynes, J.N., 1890).

The point has recently been elaborated in a context at least partly Indian by Polly Hill (1982, pp. 7-15). I am also reminded here of a wise remark by Colin Renfrew about attempts to apply Christaller to archaeology. He criticises 'mechanistic and uncritical applications' which carry with them

> all the limitations and assumptions inherent in Christaller,. . . which can be seen as one pioneering but highly specific instance of the exploration of the spatial implications of an organisational system (Renfrew,C., forthcoming).

(4) Christaller indeed stressed that he was making simplifying assumptions, for example, about uniform natural, economic and socio-political conditions.

(5) As is well-known he deduced (or claimed to have deduced) that arrangements in space would produce the most economic allocation of central places on what he called the 'marketing principle', given that any centrally provided commodity or service would have a limited range beyond which it would be uneconomic to provide it to dispersed settlements from the central places in question: a short distance for bread or doctors, a long distance for newspapers or special hospitals. Each central place thus produces a field or 'complementary region' according to the range of goods and services it provides.

(6) As is even better known, he deduced on the

marketing principle that the urban system would consist of hierarchical nesting and overlapping regular hexagonal fields.

(7) Christaller demonstrated further that the number of central places would rise in geometric progression, by a factor of three, as one went down the hierarchy: thus if there was one central place of the first order, there would be three of the second order, nine of the third order and so on. He envisaged seven orders in a stepped hierarchy. He suggested a typical population for the central places of each order and rank ranging from 5,00,000 for towns of the first order (he clearly had Munich in mind), down to a population of 1,000 for 'urban villages'.

(8) It is worthwhile mentioning his dismissal of the rank/size rule (Auerbach, F., 1913) as 'a most incredible law' and 'not much more than playing with numbers'. This rule, often mis-stated, has it, of course, that the product of rank and town population is constant; thus, if the town of the first rank has a population of 5,00,000, the towns of the second rank will have a population of 2,50,000 and so on. Indeed, Christaller's own tabulations do not support the rule.

(9) Now, clearly, to join Christallerian central places by road or rail would be a transport planner's nightmare; so he contemplated an alternative, *traffic principle* according to which the distribution of central places would be such that the maximum number of important places would lie on a single route (leading obviously to irregularity in complementary regions). He also contemplated a socio-political or 'separation' principle to cover the case where an area had a number of separate political units (like Germany before unification; or, one might postulate, India before independence, with its princely states). I propose to leave both these alternatives to the marketing principle on one side because he showed that the latter principle was usually dominant; and most authors who have followed Christaller have, in fact, forgotten his alternative principles (not always with good results).

(10) I will, however, just mention A. Lösch (1954) who basically went beyond Christaller to produce a system which would give the maximum coincidence of higher order central

places to satisfy all three of Christaller's principles; giving a pattern of differing sectors radiating from the major central places and a continuous series of centres, rather than a stepped hierarchy.

(11) I have emphasised that Christaller's theory is *static* in the sense that its importance is to explain the system of central places and their relation to complementary regions to each other at *a point in time.* In other words, it is *descriptive*—with more than an eye to southern Germany at the time when he studied it. And one might also emphasise that he was clearly thinking mainly of 'developed' countries. He did, however, amongst other things :

(a) Aver that in a densely peopled, wealthy, industrial region there would be more higher order central places than in regions with opposite characteristics (the implication being that one might develop into the other).

(b) Include a section on 'dynamic processes' which discussed the way in which a system of central places might be affected by endogenous and exogenous changes, for example, changes in population, in the supply and demand for "central goods", and in prices; and, interestingly, he foresaw the role of the motor car in dispersing population and so on. But this section is quite brief and is clearly *dynamic in a very limited and speculative sense.*

A certain amount of work by historical geographers both inside and outside Christaller's framework has indeed shown why systems of central places have changed over time. For example in the United Kingdom 18th century changes in road transport, in the 19th century in the railway network and in the 20th century in 'buses and cars, led to the obsolescence or at least decline of old, closely placed service centres well down the hierarchy which had served villagers who could only come on foot : "Malmesbury : the Decline of a Country Town", indeed.

(12) Christaller himself was very brief on policy issues

(i.e. on prescription), not surprisingly given his liberal/classical framework. In fact, having asked how the state may influence systems of central places he answers the question by saying that the state can remove 'hindrances' to development so that market forces may lead to the optimum pattern; also he says the state can choose appropriate locations, in tune with his principles for state agencies. (But I find this passage (pp. 124-26) not a little obscure, in translation, at any rate.) He does, however, say rather vaguely that 'in colonial or less developed regions' where there are 'unsuitable or imperfect systems' of central places, planning might go further and work towards the theoretical optima.

What implications do I draw out of these points that resulted from revisiting Christaller ? What, that is, for the *prescription* of policy and implementation in a planned economy, officially committed to the establishment of a 'socialist pattern of society', as in India ? First, given Christaller's ideological origins in liberal, classical, free-market economics, given too the descriptive, static—if not fundamentally inductive—nature of his theoretical model, and the brief and vague nature of his own musings on a possible role for the state, then to say the least one should think of using Christaller prescriptively with fearful caution, not least in connection with planned growth poles. Secondly, doubt must be cast on the relevance of the Christallerian hierarchy and its regular hexagonal fields to regions whose history, economy, society and physical geography differ markedly from the southern Germany of Christaller's day. Thirdly, we must note Christaller's scorn for Auerbach and the rank-size rule and expect some difficulty in prescribing an urban system that shall fulfil the requirements of the models both of Christaller and Auerbach, while recognising of course that many developing countries, including India, are deficient in middle-order central places/towns on both models.

But it is time for me to go into a little more detail in the light of my revisit to Christaller, with particular reference to India, and to those like E.A.J. Johnson, mentioned briefly earlier, who have thought in terms of the developmental desirability of setting up a full Christallerian hierarchy: in particular, of filling gaps which, as we have seen, may exist (NCAER, 1965; Johnson, 1970). Johnson and his like have been criticised by Wanmali, Barbara Harriss and others, on a number of grounds, for example :

(a) Johnson and others have been misled by census definitions, which tend to under-estimate the number of lower-order central places. Wanmali has shown that many places not classified as towns have central place functions. Thus in Singhbhum district of Bihar the ratio of census towns to villages is 1 : 271, but the ratio of market centres to villages is 1 : 23 (Wanmali, 1981, p. 191).

(b) Indeed, in many developing countries, in all continents, some central place functions are indeed undertaken not by 'urban' settlements but by periodic markets which may be located not only in towns but also, or even more perhaps, by the wayside or at crossroads or in what otherwise is clearly a village. These markets play a very important part in economic and social life. Barbara Harriss shows that in Tamil Nadu, for example, such markets are very important in the bullock trade, especially for poor people who, for caste or other reasons, may avoid the towns (Harriss, B., 1976 b). Here we have, indeed, a non-Christallerian uneven economic surface. Not all people have equal access to central places; and Wanmali has suggested prescriptively that, in order to reach the poor, some state services need to be mobile (Wanmali, 1983 a, p. 10).

(c) One underlying assumption, explicit in Johnson, is that 'modernisation', especially in the sense of technical innovation, is necessarily a city-generated phenomenon which spreads down the urban hierarchy to the poor benighted villagers. (A remark of Johnson is significant here : he refers to Mahatma Gandhi's "wholly unwarranted panegyric for fusty villages"). It follows from this assumption that an urban hierarchy must precede development. In fact, a whole school of (significantly) American geographers like E.W. Soja talk of the geography of modernisation in precisely those terms (e.g. Soja, E.W., and Tobin, R.J., 1972). But, I would submit, this is all very Americocentric. And while some innovations, desirable or even necessary for rural development, did so flow, or at any rate start, in *some* towns, other city bred innovations are disastrously irrelevant (e.g. the modern ricemill that rouses Barbara Harriss' criticism (Harriss, B., p. 77, pp. 276-99) : while *relevant* innovations, like mechanised or electrified irrigation and high yielding varieties may largely come by other routes; the extension services, the radio, the vernacular press, conversation in all sorts of places—that is, by non-hierarchical diffusion. Soja also thought, incidentally, that colonial central places serve colonial masters and need replacing in independent developing countries.

(d) There is also a conflation, if not confusion (especially on the part of E.A.J. Johnson, aforesaid (1970)) of what, as I have emphasised is the descriptive and essentially static Christaller theory with the essentially dynamic and prescriptive growth pole theory and policy stemming ultimately from the French economist F. Perroux. These rest on the idea that planned spatial aggregation, particularly of industry, can attract other industries and play a part in regional planning strategies and in hierarchical diffusion as just discussed. I do not here criticise the growth pole idea, though I am basically sympathetic to Barbara Harriss' strictures (Harriss, B., 1976a). And there is, I suppose sufficient empirical evidence that growth poles, planned or unplanned can work *in time* : compare, for example, the aggregation of industry round the 70-year old steel works of Jamshedpur and the lack of such aggregation so far round later Indian steel work towns (though other factors may be at work here); and note also the way in which such industrial towns, "Boulders in a plain" indeed, have become central places in time at the expense of earlier central places. This has happened with Jamshedpur, as Wanmali has shown (Wanmali, 1981) : and with Swindon, some time after the establishment of the railway works there in the 19th century (hence its successful competition with Malmesbury). But I do not criticise the idea that growth poles have to fill gaps, above all gaps in the Christaller or Auerbach schemes. Indeed, there is no logical relationship whatsoever between central place theory, essentially a theory of urban/regional relations; and growth pole theory, which has to do with industrial aggregation. Christaller, indeed, has been criticised (e.g., by R.E. Dickinson 1934, p. 44) on the grounds of his *neglect* of industrial towns. So let us not confuse central places and *industrial* growth poles.

(e) At least the growth-pole men have a way of operationalising town growth. Other exponents of the dubious art of making Christaller dynamic, of filling gaps in the hierarchy, of fulfilling the rank/size rule have not. In fact, they really have no answer to the questions "How does one provide central places of the right size at the right spacing and with the right functions?"; and "How do we know any of these parameters given the historical fact that size, spacing and function can be shown to have varied over time where it is possible to document the evolution of an urban system?". Even if the answer comes back that such provision is desirable, it must rest on a proper study of all kinds of central places, permanent and temporary, from the village. And, I submit, on the doctrine that rural places, whatever else they may or may not need, do demand rural solutions to their problems. I believe too, with Mark Jefferson (1931), that "Cities do not grow up of themselves (I would add, nor can they be created). Countrysides set them up to do tasks that must be performed in central places". Barbara Harriss put it more pithily by saying that economies "get the towns they deserve" (Harriss, B., 1980).

(f) There is, however, this to be said to soften the rather hard line of the preceding paragraph. One thing that I have learnt from my contact with Sudhir Wanmali is that in an interventionist state like India there is a good deal that governments can do to strengthen central places. A very potent way, as Wanmali's work in Miryalguda Taluka of Nalgonda District, Andhra Pradesh shows, is to provide a successful irrigation scheme, in this case the canals flowing from Nagarjunasagar on the river Krishna, in an area that when Wanmali first studied was dry and, indeed, drought-prone (Wanmali, 1983a and 1983b). Here, indeed, the countryside has set up central places to undertake many tasks from the marketing of the new-found agricultural surplus to the provision of retail services that mop up some of the newly created wealth. But the state has also, largely though not exclusively with an eye to enabling farmers to take advantage of the potentialities linked to irrigation, established a number of rural services, and located them in fixed settlements which thus become strengthened central places, or even new central places. Between them, then the government and the private traders and shopkeepers have answered the question in the preceding paragraph "How does one provide (or at any rate strengthen) central places". But there was, and need not be any appeal to Christaller (except, perhaps, for his reference to the state's location of its agencies)—or to Auerbach ; and assuredly no case for conflating the two, or mistaking the statement about size of the second for the statement about function of the (*cf.* Alam, S. Manzoor and Khan, W., 1972 and Reddy, N.B.K., 1975)—though it may be that over time a more nearly Christallerian hierarchy may emerge, as P. Haggett suggested it might be in the area round the new capital city of Brasilia, so that state organisations may be able to predict future needs for urban planning by assuming that it will (Haggett, P., 1972, pp. 286-87). But, I would submit, this is less likely while India remains an interventionist state, given the quintessentially free-market assumptions underlying both the Christaller model and Haggett's suggestion (for shifting of policies and deficiencies of implementation may distort what would happen if these assumptions were fulfilled) ; and given, too, another point that I have laboured : that for all his protestations to the contrary Christaller's model is basically inductive and derived from a situation that reflects the history and society of southern Germany.

There may be some readers, if any can be persuaded to read so far, who may feel that all this is very eccentric. But I derive comfort, and hope that such readers may be stimulated to think, by two facts ; that my views on growth poles in relation to 'modernisation' find substantial confirmation in John Friedmann and Clyde Weaver's book *Territory and Function : the Evolution of Regional Planning* (1979), which represents a significant shift from Friedmann's earlier views (though for all my rural bias I do not go wholly along with them in their notions on "agropolitan development", particularly so far as local autarky is concerned), and in K. V. Sundaram's most recent work (1983, ch. 4) : and that in his two most recent works Sudhir Wanmali shares very much my views about the undesirability of trying to create a Christallerian hierarchy in a strange environment and of conflating and confusing central places and growth poles, about the need to view the scene from the village, and about the lamentable deficiencies of any attempt to model development as the diffusion of "modernisation" simplistically conceived, (1983a, 1983b).

REFERENCES

Alam, S. Manzoor and Khan, W. (1972). *Metropolitan Hyderabad and its Region.*

Auerbach, F. (1913). Das Geset der BeVolkerungskonzentration, *Petermanns Mittellungen*, 59, 74.

Bracey, H.E. (1952). *Social Provision in Rural Wiltshire*, London.

Brookfield, H.C. (1975). *Interdependent Development*, London.

Chisholm, M. (1979). *Rural Settlement and Land Use*, 3rd edn., London.

Christaller, W. (1933). *Die zentralen Orte in Süddeutschland : Eine ökonomisch—geographische Untersuchung über die Gesetzmässigkeit der Verbreitung und Entwicklung der Siedlungen mit städtischen Functionen*, Jena.

Christaller, W., trans. Baskin, C.J. (1966). *Central Places in Southern Germany*, Englewood Cliffs, N.J.

Dickinson, R.E. (1932). The distribution of functions of the smaller urban settlements of East Anglia, *Geography*, 17, pp. 19-31.

Dickinson, R.E. (1934). The markets and market areas of East Anglia, *Economic Geography*, 10, pp. 172-82.

Farmer, B.H. (1974). *Agricultural Colonization in India since Independence*, London.

Febvre, L., trans. Mountford, E.G. and Paxton, J.H. (1925). *A Geographical Introduction to History*, London.

Friedmann, J. and Weaver, C. (1979). *Territory and Function : The Evolution of Regional Planning*, London.

Haggett, P. (1972). *Geography : A Modern Synthesis*, New York.

Harriss, B. (1976a) The Indian ideology of growth centres, *Area* 8, pp. 263-9.

Harriss, B. (1976b). Social specificity in rural weekly markets.—The case of Tamil Nadu, *Mainzer, Geographischen Studien*, 10, pp. 39-48.

Harriss, B. (1977). Paddy-milling : Problems in policy and the choice of technology, Ch. 19 in Farmer, B.H. (ed.), *Green Revolution ?*

Technology and Change in Rice-growing Areas of Tamil Nadu and Sri Lanka, London.

Hermansen, T. (1972). Development Poles and Related Theories : A Synoptic Review, pp. 160-203, in Hansen, N.M. (ed.), *Growth Centres in Regional Economic Development*, New York.

Hermansen, T. (1975). Spatial Organization and Economic Development : The Scope and Tasks of Spatial Planning in Kuklinski, A. (ed.), *National Policies and Plans*, Geneva.

Hill, P. (1982). *Dry Grain Farming Families*, (Cambridge).

Hirschman, A.O. (1958). *The Strategy of Economic Development*, New Haven, Conn.

Johnson, E.A.J. (1970). *The Organization of Space in Developing Countries*, Cambridge, Mass.

Keeble, D.E. (1967). Models of Economic Development, in Chapter 8 *in* Chorley, R.J. & Haggett, P., *Models in Geography*, (London).

Keynes, J.N. (1890). *The Scope and Method of Political Economy*, London.

Lipton, M. (1977). *Why Poor People Stay Poor*, London.

Losch, A. (1954). *The Economics of Location*, New Haven, Conn.

Mabogunje, A. (1980). *The Development Process : A Spatial Perspective*, London.

Misra, R.P., Prakasa Rao, V.L.S. and Sundaram, K.V. (1970). *Growth Poles and Growth Centres for Regional Economic Development in India*, Geneva.

Misra, R.P., Sundaram, K.V. and Prakasa Rao, V.L.S. (1974). *Regional Development Planning in India : a New Strategy*, (New Delhi).

Myrdal, G. (1957). *Economic Theory and Under-developed Regions* London.

NCAER (National Council for Applied Economic Research) (1965). *Market Towns and Spatial Development in India*, New Delhi.

Reddy, N.B.K. (1975). Rank Size Distribution of Towns in Rayalaseema, Paper presented to the First South East Asian Commonwealth Geographical Congress, Singapore.

Renfrew, C. (forthcoming). Contribution to Symposium on Geography, Archaeology and Environment, *Geographical Journal* (forthcoming).

Sen, Lalit K., Wanmali, S. *et al.* (1971). *Planning Rural Growth Centres for Integrated Area Development*, Hyderabad.

Smailes, A. (1944). The urban hierarchy in England and Wales, *Geography*, 29, pp. 41-51.

Smailes, A. (1947). The Analysis and Delimitation of Urban Fields, *Geography*, 32, pp. 151-61.

Soja, E.W. and Tobin, R.J. (1972). The Geography of Modernization : Paths, Patterns and Processes of Spatial Change in Developing Countries, in Brunner, R. and Brewer, G. (eds.), *Ordered Complexity : Empirical Theories of Political Development*, New York.

Sundaram, K.V. (1983). *Geography of Underdevelopment*, New Delhi.

UN (United Nations) (1976). *Habitat : UN Conference on Human Settlements, Vancouver, 1976.* New York.

Wanmali, S. (1973). Spatial Behavioural Pattern of Urban-Rural Interaction, *Behavioural Sciences and Community Development*, 7, pp. 17-30.

Wanmali, S. (1981). *Periodic Markets and Rural Development in India*, (Delhi).

Wanmali, S. (1983a). *Service Provision and Rural Development in India : A Study of Miryalguda Taluka*, Research Report 37, International Food Policy Research Institute, Washington, D.C..

Wanmali, S. (1983b), *Service Centres in Rural India*, Delhi.

11

IMPACT OF CENTRAL PLACE THEORY ON SETTLEMENT GEOGRAPHY OF INDIA

K. SITA AND V.S. PHADKE

Introduction

FIFTY years have elapsed since Christaller propounded the Central Place Theory. This pioneering work on the theoretical aspects of settlement geography has had an unparalleled impact on research work in geography all over the world. It has generated numerous studies on both rural and urban settlements, and even today, though its hexagonal geometry is not accepted, the concept of functional organisation and its logic continue to provide the theoretical basis for studies of settlements. It seems appropriate, on this occasion, in view of Dr. Prakasa Rao's keen interest in and contributions to this field[52], [53], [54], [55] to review the impact of the theory on studies of Indian settlements. The works of Berry and Pred[6] and later authors[4] have indicated the main lines of work in western geography, but lack an adequate coverage of Indian work. It would be presumptuous to claim that this paper has exhaustively covered all the work in this field ; the authors have tried to identify the major trends of research in Indian settlement geography which appear to have emanated from the central place theory.

THE MAJOR AREAS OF APPLICATION OF CENTRAL PLACE THEORY

The major area in settlement geography where the early impact of the central place theory was apparent was in the understanding of functional hierarchies of settlements ; these have been derived for a number of regions. Most of these studies are at the micro-level and there is a lack of standdardised methodology[14, 55]. Studies in verification or establishment of the aspects of the theory related to range of goods, thresholds, shape of complementary regions are relatively few. In recent years numerous studies have been carried out in which the central place theory has been applied for understanding the existing functional organisation in space, identifying the spatial gaps, and modifying the structure for integrated area development. In these studies, the central place theory provides a basis for identification of growth centres. It may be noted that there are a few studies[7, 42, 61, 75] which try to fit the central place theory into general systems theory and in the process have integrated it with other concepts such as spatial interaction.

STUDIES IN SETTLEMENT HIERARCHIES

It is difficult to distinguish studies on centrality from studies on hierarchic arrangement of settlements as most researchers have used the centrality scores for identifying the hierarchy of settlements in their study area. However, in some studies, the emphasis is on developing a methodology for deriving centrality. Wanmali[105, 107], after a survey of methods to determine centrality in different countries, argues that all functions should be taken into account. He outlines the central functions and the weightages to be assigned to each, and illustrates it with the example of the Nagpur Metropolitan Region. Deshmukh[19] attempts to find out to what extent population differences account for functional status of settlements in Buldana district. Singh[78] has derived two types of centrality indices, the relative centrality index and the absolute centrality index for determining the centrality of service centres in Uttar Pradesh. Diddee and Dikshit[21] have argued that for determining centrality of small and medium sized central places, the surplus function method is most suitable and have applied it to Pune district. Sawant and Bhole[65] have used the same method in their study of the Indrayani basin. Biswas[13] bases his centrality index on excess of secondary and tertiary activities in towns. Sharma[71] identifies centrality based on concentration of working population, while Singh and Singh[84] use concentration of commercial population.

Centrality indices based on various parameters relating to transport and accessibility have been important in Indian studies due to less well developed transport which has resulted in accessibility playing a significant role in the emergence of centrality[15, 23, 25, 32].

Sophisticated techniques such as scalogram and factor analysis form the basis of some studies[9, 14, 62, 66, 68, 70, 99]. For example, Sharma[68] applies the Principal Component analysis for identifying the ranks of settlements in Karnal district. In all these studies, the hierarchy of settlements is derived after classification of the centrality scores.

There are numerous studies on hierarchy of settlements either in administrative units such as states, districts or in geographical units. For Mysore, a pioneering work is that of Prakasa Rao[52], while Vishwanath[103] has considered the small towns. Gopalakrishnan[29] has identified five hierarchical types in Tamil Nadu while Barai[3] has selected two talukas of Tamil Nadu and arrived at a four-fold hierarchy. Sinha and Mandal[91] have identified the hierarchy of trade towns in Bihar, Mandal[41] and Thakur[100] for Bihar plain, while Bronger[14] identifies the hierarchical classes in Andhra Pradesh. Singh[79, 80], based on his study of major central places in Uttar Pradesh, finds that the hierarchical grading systems are not satisfactory. Bhattacharya[12] identifies six hierarchical orders in North Bengal and found that administrative status was the most important factor in determining a town's hierarchical position. He has also[11] derived six hierarchical orders in Darjeeling district and has attempted to correlate the hierarchy with the genetic types. Jain[33] has attempted to establish the urban hierarchy in Vidarbha on the basis of telephone services.

Singh *et al.*[85] have identified the service centres of Baghelkhand on the basis of 'centrality' and 'functionality' and conclude that a K-12 network for higher order centres and K-3 for lower order centres would be suitable. A similar methodology is used by Singh *et al.*[86] in the analysis of the spatial system of towns of Dandakaranya, and by Iyer and Jain[30] in the study of central places of Purna valley. Sita[92] compares the size hierarchy with the functional hierarchy of settlements in different parts of South Konkan. Deshpande *et al.*[20] have worked out the hierarchy in South Kolaba (Raigad), an area in the shadow of Greater Bombay, in order to locate the functional gaps. Verma Singh[103] derive a nodality index for grading central places in Madhubani and compare the empirical results with the theoretical postulates. Jafri[31] and Sinha[89] use deviation from the mean for deriving the hierarchical orders, while Singh[87] recognises four levels of hierarchy in the upper Champavati Basin. Singh[76] has analysed the spatial pattern of central places in the Middle Ganga valley. Thaha[98] recognises four hierarchical levels in Alur block of Kurnool district.

Metropolitan regions being functional and planning units have attracted studies aimed at identifying the present spatial organisation of settlements which would throw light on the functional saps. Swaminathan[96] has investigated the hierarchy of settlements in the Coimbatore Metropolitan Region using multivariate techniques. Kar[35] has highlighted the urban hierarchy around Calcutta while Nangia's study[50] focuses on the Delhi Metropolitan Region. Kulkarni[36] has worked on the hierarchy for the Nasik Metropolitan Region. Sita and Phadke[93] have analysed the similarities and differences between the size and functional hierarchy of settlements in the Bombay Metropolitan Region.

STUDIES ON VERIFICATION OF CONCEPTS

Concept of threshold

Rushton[64] has cautioned that the technique for determining population thresholds used in western studies is not applicable in the Indian context as there may not be constant multiplier effects between the population of the tributary area and the size of the centres. This viewpoint that Indian conditions require modifications is also stressed by Ensminger and Roy[24]. Gopalakrishnan[29] in his study of small towns of Tamil Nadu has determined the threshold population for services to identify the standard services. Singh[87] has adopted the Reed-Muench method for determining thresholds of functions while Misra[48] has considered the minimum population of a settlement in which a function exists as indicating the threshold.

Range of goods

The demonstration of the operation of the range of central goods had been made by Mayfield[43, 44] in his work on Punjab. Prakasa Rao and Ramachandran[56] based on a sample survey of households have focused attention on the differences in spatial preference not only between goods and services but also between the rich and the poor. Wanmali[108] and Ghosal[28] bring out the differences between the actual pattern of tributary areas and the theoretical expectation.

Shape of complementary regions

A few studies have attempted to verify the shape of the complementary regions. Wanmali and Khan[109] have stressed the departures from the classic hexagonal model. Prakasa Rao[53] has drawn attention to the departure from hexagonal market areas in Muzaffarnagar district where rectilinear rhomboidal trade areas give a better fit. Sharma[72] points out that the central places in the Indian desert operate in elongated arc-shaped areas due to the impact of transport and communication.

Spacing of central places

The spacing of central places has not received as much attention as their hierarchy due to the difficulty of comparing a distributional pattern on a non-isotropic surface with the theoretical pattern on an isotropic surface. Prasad[58] and Singh[80] compare the actual distributions of central places with the hypothetical one in their study areas. Kumar and

Sharma[38] focus attention on spatial arrangements of urban centres while Varma[101] examines the distance-size and the distance-function relationship in the urban settlement structure of Kota district. Singh and Singh[84] have made use of a more sophisticated index that is the 'Convergent Ratio' to study distributional patterns. Dayal and Thakur[16] apply the entropy method to show that the urban settlements of Bihar are tending to be more uniformly distributed.

CENTRAL PLACE THEORY AND REGIONAL PLANNING

One of the applications of central place theory (CPT) is an analysis of the functional organisation of space and in consideration of central places as potential growth centres for the planning of integrated development of areas [26], [27], [63], [67]. The initial application of this idea is in some of the master plans for regional development[17], [34], Singh[83] though more concerned with the distributional pattern of settlements in Varanasi, makes a reference to locating services at selected villages for community development.

Attempts to study the functional and spatial hierarchy, identify gaps and suggest idealised functional spatial patterns are more recent and the international seminar on the 'Kanpur Region'[18] appears to have provided a triggering effect. Suggestions for development of the Kanpur Region, in which the existing administrative and commercial centres would play a role have been put forward[5], [8], [18]. Wanmali[104] attempts to identify the developmental dualism between the Nagpur region and the rest of Vidarbha and suggests a four-tier structure for effective functional organisation. He is of the opinion that these service centres could effectively transmit developmental impulses to the smallest units. Later studies[1], [66], [69], [73] suggest functional organisation for spatial integration of regional economics. The central places are not looked upon as mere 'service centres' as there is an emphasis on growth[59]. Bhat[9] considers the settlement hierarchy derived through factor analysis and composite scores and relates it to the integrated development of the area. Bronger[14] emphasises the need for proper coordination of central places at various levels if they are to act as development foci, and the exercise carried out for Jangaon taluka in Andhra Pradesh by Reddy, *et al.*[60] suggests a rational spatial and functional framework of settlements. Singh[81] and Singh, *et al.*[82] in their studies on the Central Plains of Uttar Pradesh suggest a model for the planned growth of settlements in the area; they recommend a K-4 network for the area. Singh and Singh[84] have focused attention on variations in the characteristics and service efficiency of settlement system in different spatial contexts, and indicate their importance in micro-level planning. This is suggested by a number of case studies.

Some workers have attempted the application of CPT for spatial planning in special regional contexts. Bhat, *et al.*[10] have derived the central place hierarchy for the developing agricultural region of Karnal area while Sundaram[94] stressed the evolution of the settlement hierarchy for the irrigated colonies in the Rajasthan Canal Area. Minocha[46] and Singh[74] focus attention on spatial organisation in backward areas, while Sundaram, *et al.*[95] and Kumar and Sharma[37] deal with spatial planning in tribal areas.

CENTRAL PLACE THEORY AND GENERAL SYSTEM THEORY

Mallick[39] discusses the significance of CPT for providing an overall framework of settlement system and for identifying growth and service centres, while Bhat[7] provides a conceptual spatial framework for Regional and National Planning in India. Mandal[42] draws attention to the orientation of the transportation links to the next higher level centres, suggests the location of urban centres at different levels in the Loschian landscape and stresses the need for road planning. Singh and Ved Prakash[75] make a critical analysis of 'central place' and 'spatial interaction' notions with a view to synthesise them in the framework of general systems theory. The difference between horizontal and vertical interactions is brought out. Roy[61] instead of looking at growth centres as foci for development of the surrounding area emphasises the evolution of a system of regions based on potentiality which guide the pattern of growth of service centres.

GAPS IN RESEARCH WORK

The aspect of central place theory which has evoked a plethora of work is that of centrality and hierarchy of settlements. However, the studies have proceeded on individual and sometimes mechanical lines. It is difficult to identify a common denominator and as such studies lack comparability and are difficult to apply in planning. Though local and regional differences are bound to exist, it would be desirable to evolve a more standardised framework suited to Indian conditions in the light of the existing literature.

The studies related to empirical testing of the postulates of the theory have been very few. There is need for a larger number of such studies. In India, where the information level is low, and social and psychological factors affect consumer space preference, there are considerable possibilities of behavioural modifications of the basic economic tendencies. Identification of these would go a long way in making growth centres more effective in transmission of developmental impulses.

The usefulness of central place theory in micro-level planning has been proved beyond doubt. The main problem in India appears to be a lack of communication between the administrators on the one hand, and the academic community on the other. The results of academic work, which could be usefully incorporated in planning better location of facilities have not been given due weightage.

REFERENCES

1. Alam, S.M. and Khan, W. (1972). *Metropolitan Hyderabad and its Region—A Strategy for Development*, Asia Publishing House, Bombay.
2. Banerji, S. and Fischer, H.B. (1974). Hierarchical location analysis for integrated area planning in rural India, *Papers of the Reg. Sc. Assoc.* 33, 177-94.
3. Barai, D.C. (1970). Hierarchy of settlements in Tamil Nadu, *I.G.J.*, 45 (3 and 4), 65-72.
4. Beavon, K.S.O. (1977). *Central Place Theory : A Reinterpretation*, Longman, New York.

5. Berry, B.J.L. (1969). Policy implications of an urban location model for the Kanpur Region in P.B. Desai *et al.* (ed.), *Regional Perspectives of Industrial and Urban Growth : The Case of Kanpur*, Macmillan, Bombay, 203-19.
6. Berry, B.J.L. and Pred. A. (1961). *Central Place Studies : A Bibliography of Theory and Applications*, (Regional Science Research Institute, Bibliographic Series 1).
7. Bhat, L.S. (1969). Central place model as a spatial framework for regional and national planning in India, *Symposium on the City as Centre of Change in Asia*, University of Hongkong.
8. Bhat, L.S. (1969). A regional approach to the urban and industrial development of the Kanpur Region, in P.B. Desai *et al.* (ed.), *Regional Perspectives of Industrial and Urban Growth : The Case of Kanpur*, Macmillan, Bombay, 230-40.
9. Bhat, L.S. (1976). *Micro-Level Planning : A Case Study of Karnal Area (Haryana)*, K.B. Publications, New Delhi.
10. Bhat, L.S., *et al.* (1974). *Central Place Hierarchy for a Developing Agricultural Region : A Study of Karnal Area*, Report of the Study sponsored by ICSSR, New Delhi.
11. Bhattacharya, B. (1971). A genetic classification of settlements and their hierarchical order in the district of Darjeeling, *G.R.I.*, 33(3), 183-95.
12. Bhattacharya, B. (1972). Factors determining the central functions and urban hierarchy in North Bengal, *G.R.I.*, 34(4), 327-38.
13. Biswas, S.K. (1982). Quantitative evaluation of the development potentialities of the towns of Bijnor district, in M. Alam and K.N. Gopi (ed.), *Settlement System of India*. Oxford and IBH, 87-92.
14. Bronger, (1978). Central place system, regional planning and development in developing countries : A case of India", in R.L. Singh and R.P.B. Singh (ed.), *Transformation of Rural Habitat in Indian Perspective : A Geographical Dimension*, NGSI Varanasi, 134-64.
15. Chandra, G. (1972). Accessibility in relation to settlement in Ghar (Siwalik Zone), district Saharanpur, *Proc., Ind. Geog. Cong.*, 13-14.
16. Dayal, P. and Thakur, B. (1982). Settlement systems of Bihar : The dynamics of urban settlement systems, in M. Alam and K.N. Gopi (ed.), *Settlement System of India*, Oxford & IBH, 144-67.
17. Delhi Development Authority (1961). *Master Plan for Delhi*, Delhi.
18. Desai, P.B., *et al.* (ed.) (1969). *Regional Perspectives of Industrial and Urban Growth : The Case of Kanpur.* Papers and Proc. of the Kanpur International Seminar, 1967, Macmillan, Bombay.

19. Deshmukh, V.M. (1970). Correlation between size of population and social provision in the settlements of the Buldana district, *N.G.J.I.* 16(1), 79-83.

20. Deshpande, C.D., *et al.* (1980). *Impact of a Metropolitan City on the Surrounding Region*, Concept, New Delhi.

21. Diddee, J. and Dikshit, K.R. (1979). A note on measuring the centrality of small and medium size central places, *I.I.G.*, 1(1), 69-78.

22. Dutt, A.K. (1969). Intra-city hierarchy of central places: Calcutta as a case study, *The Prof. Geog.*, 21(1), 18-22.

23. Dutt, A.K. and Banerjee, S.B. (1970). Transport index in West Bengal: A means to determining central place hierarchy, *N.G.J.I.*, 16 (3 & 4), 199-207.

24. Ensminger, D. and Roy, P. (1972). Growth centres and viable rural-urban communities, in L.K. Sen, (ed.), *Readings in Micro-Level Planning and Rural Growth Centres*, NICD, Hyderabad, 17-21.

25. Folke, S. (1968). Central place systems and spatial interaction in Nilgiris and Coorg (South India), *21st Int. Geog. Cong.*

26. Ford Foundation (1973). *A Geographical Approach to Settlement Planning for Integrated Area Development*, Ford Foundation, New Delhi.

27. Ford Foundation (1973). *The Pilot Research Project in Growth Centres*, 3rd Progress Report, Ford Foundation, New Delhi.

28. Ghosal, K. (1972). Market places and market areas, *Nat. Geogr.* 7, 85-95.

29. Gopalkrishnan, K.S. (1979). Hierarchical classification of small towns in Tamilnadu, in R.L. Singh and R.P.B. Singh (ed.), *Place of Small Towns in India*, NGSI Res. Pub. 21, 113-18.

30. Iyer, K.V. and Jain, N.C. (1979). Hierarchy of central places in Purna valley, *Dec. Geog.* 17(2), 590-604.

31. Jafri, S.S.A. (1972). Hierarchy of service centres—A case study of four hill districts of U.P., *Proc., Ind. Geog. Congress*, 10-13.

32. Jain, J. (1982). Spatial structure of urban settlement system of Malwa plateau, M.P., in M. Alam and K.N. Gopi (ed.) *Settlement System of India*, Oxford & IBH, 122-43.

33. Jain, N.G. (1971). Urban hierarchy and telephone service in Vidarbha (Maharashtra), *N.G.J.I* . 17(2 & 3), 134-37.

34. Joint Committee for Diagnostic Survey of the Damodar Valley Region (1968). *Damodar Valley Survey Reports and the Planning Atlas of the Damodar Valley Region*, Calcutta.

35. Kar, N.R. (1960). Urban hierarchy and central functions around the city of Calcutta and its significance, *Proc., I.G.U. Symposium on Urban Geog.*, Lund, Sweden.

36. Kulkarni, K.M. (1981). *Urban Structure and Interaction : A Study of Nasik City Region*, Concept, New Delhi.

37. Kumar, A. and Sharma, N. (1973). Rural service centres in tribal setting, *Proc. Settlement. Geog Conf.*
38. Kumar, A. and Sharma, N. (1974). Spatial arrangements and degree of specialization of urban centres, *G.R.I.* 36(2), 120-30.
39. Mallick, U.C. (1979). Development of small towns for balanced urbanization and economic growth, in R.L. Singh and R.P.B. Singh (ed.), *Place of Small Towns in India,* NGSI, Res. Pub. 21, 55-67.
40. Mandal, R.B. (1973). Rurban centres in Bihar plain, *Ind. Geog. Studies*, 1, 12-23.
41. Mandal, R.B. (1975). Central Place hierarchy in Bihar plain, *N.G.J.I.*, 21(2), 120-34.
42. Mandal, R.B. (1979). Law of village and rurban centres : Location on Loschian landscape, in R.L. Singh and R.P.B. Singh (ed.), *Place of Small Towns in India,* N.G.S.I. Res. Pub. 21, 79-83.
43. Mayfield, R.C. (1960). Conformations of service and retail activities, *Lund Symposium on Problems of Urban Geog.*, 19th Int. Geog. Cong.
44. Mayfield, R.C. (1963). The range of a central good in the Indian Punjab, *AAG*, 53(1), 38-49.
45. Mayfield, R.C. (1967). A central place hierarchy in Northern India, in W.L. Garrison and D.F. Marble (ed.), *Quantitative Geography, Part I : Economic and Cultural Topics.* (Dept. of Geog., Northwestern University, Evanston, Illinois).
46. Minocha, A.C. (1974). Planning for social services in a backward region, *Ind. Jl. of Reg. Sc.*, 6(2), 181-98.
47. Misra, G.K. (1972). A methodology for identifying service centres in rural areas—A study in Miryalguda Taluka, *Journal of Behavioural Sciences and Community Development*, 6 (1), 89-97.
48. Misra, G.K. (1979). Role of small towns in district planning : A case of Raichur District, in R.L. Singh and R.P.B. Singh (ed.), *Place of Small Towns in India*, NGSI Res. Pub. 21, 201-10.
49. Misra, R.P. (1969). *Regional Planning—Concepts, Techniques, Policies and Case Studies,* Prasaranga, University of Mysore, Mysore.
50. Nangia, S. (1972). Hierarchy of settlements in the Delhi Metropolitan Region, *Proc., Ind. Geog. Cong.*, 8-9.
51. Pal, M.N., *et al.* (1968). A technique for ranking central places and determining linkages, *Symposium on Quantitative Methods in Geog.*, 21st Int. Geog. Cong., Mysore.
52. Prakasa Rao, V.L.S. (1964). *Towns of Mysore State,* (Indian Statistical Institute, Calcutta).
53. Prakasa Rao, V.L.S. (1972). Central place theory, in Sen, L.K. (ed.), *Readings in Micro-level Planning and Rural Growth Centres,* (NICD, Hyderabad), 173-76.

54. Prakasa Rao, V.L.S. (1981). Central place theory in retrospect and prospect, *Annals NAGI*, 1 (2), 39-50.
55. Prakasa Rao, V.L.S. (1983). *Urbanization in India*, (Concept, New Delhi).
56. Prakasa Rao, V.L.S. and Ramachandran, R. (1971). Mobility and choice of central places, *Analytical Geography*, 1(1), 17-33.
57. Prasad, H. and Maheshwari, S.K. (1972). Hierarchy of service centres in the Parwa Dun, Dehra Dun District, *Geog. Knowledge*, 5, 1-14.
58. Prasad, L. (1979). Spatial pattern of rural central places, Patna, *Uttar Bharat Bhoogol Patrika*, 15(2), 107-14.
59. Ramachandran, R. (1976). Identification of growth centres and growth points in south-east resource region, *NGJI*, 22(1 & 2), 15-24.
60. Reddy, K.S. and Reddy, K.V. (1979). Potential growth centres—Jangaon—A case study, *Decc. Geog.*, 17(3) 716-21.
61. Roy, B.K. (1971). Potentiality of environs of growth centres in India, *IGJ.*, 46. 60-64.
62. Roy, P. and Patil, B.R. (1977). *Manual for Block Level Planning*, Macmillan, Bombay.
63. Rushton, G. (1972), Some issues in central place theory relevant to regional plan development, in L.K. Sen (ed.) *Readings in Micro-Level Planning and Rural Growth* Centres, (NICD, Hyderabad) 210-12.
64. Rushton, G. (1972). Planning practices in the context of location theories, in L.K. Sen (ed.), *Readings in Micro-Level Planning and Rural Growth Centres*, (NICD, Hyderabad), 213-19.
65. Sawant, S.B., and Bhole, A.S. (1980). Distribution, centrality and hierarchy of central places in the Indrayani Basin, *The Deccan Geog.*, 18(2), 822-34.
66. Sen, L.K., *et al.* (1971). *Planning Rural Growth Centres for Integrated Area Development*, NICD, Hyderabad.
67. Sen, L.K. (1972). A new strategy for community development: Planning for integrated area development and rural growth centres, in L K. Sen (ed.), *Readings in Micro-Level Planning and Rural Growth Centres*, NICD, Hyderabad, 27-36.
68. Sharma, A.N. (1972). Ranking of settlements and identification of spatial linkages in Karnal district, *Proc., Ind. Geog. Cong.*, 25-26.
69. Sharma, A.N. (1973). *Settlement Structure and Space Relations: A Study in Functional Organization in Karnal District*, M. Phil Thesis, JNU, New Delhi.
70. Sharma, A.N. and Bhat, L.S. (1974), Functional spatial organization of human settlements for integrated area study, *Proc. and Papers*, 13th Indian Econometrics Conference, Ahmedabad.
71. Sharma, B.L. (1975). Hierarchy of central places in the Ajmer Region, *Geog. Viewpoint*, 5 (1 & 2).

72. Sharma, R.C. (1972). *Settlement Geography of the Indian Desert*, Kumar Brothers, New Delhi.
73. Singh, J. (1972). Regional planning for rural development, in R.P. Misra (ed.), *Urban Systems and Rural Development*, Prasaranga, University of Mysore, Mysore.
74. Singh, J. (1979). *Central Places and Spatial Organization in a Backward Economy, Gorakhpur Region : A Study in Integrated Regional Development*, Uttar Bharat Bhoogol Parishad, Gorakhpur.
75. Singh, J. and Ved Prakash (1973). Central places and spatial interaction : A critical approach, *N.G.J.I.* 19 (3 & 4), 158-76.
76. Singh, K.N. (1966). Spatial pattern of central places in the middle Ganga valley, India, *N.G.J.I.*, 12, 213-26.
77. Singh, O.P. (1968). Functions and functional classes of central places in U.P., *N.G.J.I.*, 14 (2 & 3), 83-127.
78. Singh, O.P. (1971). Towards determining the hierarchy of service centres : A methodology for central place studies, *N.G.J.I.*, 17 (4), 165-77.
79. Singh, O.P. (1971). Relationships of rank, size and distribution of central places in U.P., *National Geographer*, 6, 19-30.
80. Singh, O.P. (1972). Spatial distribution of sizeable central places of U.P. on the nearest neighbour method, *National Geographer*, 7, 79-84.
81. Singh, O.P. (1975). Central place regions for planning in Uttar Pradesh, *N.G.J.I.*, 21 (2), 96-107.
82. Singh, O., *et al.* (1979). Towards planning network of small towns in the central planning of U.P., in R.L. Singh and R.P.B. Singh (ed.), *Place of Small Towns in India*, N.G.S.I. Res. Pub. 21, 193-200.
83. Singh, R.B. (1969). Rural settlement types and their distributions : Examples from Varanasi District, India, *N.G.J.I.*, 15 (2), 91-104.
84. Singh, R.L. and Singh, R.P.B. (1978). *Transformation of Rural Habitat in Indian Perspective, A Geographical Dimension*, N.G.S.I., Varanasi.
85. Singh, R.L., *et al.* (1979). Spatial analysis of small towns in Baghelkhand : Some rationales, in R.L. Singh and R.P.B. Singh (ed.), *Place of Small Towns in India*, NGSI, Varanasi, 25-32.
86. Singh, R.P.B. and Singh, G.S. (1979). Spatial system of towns of Dandakaranya : A study of status, shape and structure, in R.L. Singh and R.P.B. Singh (ed.), *Place of Small Towns in India*, NGSI, Varanasi, 33-54.
87. Singh, R.Y. (1972). Hierarchy of rural settlements and identification of functional gaps in a part of the Upper Champavati Basin, M.P., *Proc., Ind. Geog., Cong.*, 15-16.
88. Singh, S.M. (1955). The stability theory of rural central place development, *N.G.J.I.*; 2 (1), 13-21.

89. Sinha, V.N.P. (1973). Hierarchy of urban settlements in the Chotanagpur plateau, *Geog. View Point*, 4 (1 & 2), 41-55.

90. Sinha, V.N.P. (1972). Hierarchy of urban settlements in the Chotanagpur plateau, *Geog. Knowledge*, 5, 28-40.

91. Sinha, V.N.P. and Mandal, R.B. (1974). Hierarchy of trade in Bihar. *Ind. Geog. Studies*, 3, 79-84.

92. Sita, K. (1982). The hierarchy of settlements in south Konkan, *Ind. Geog. Studies*, 19, 1-8.

93. Sita, K., and Phadke, V.S. (1982). Settlement system of the Bombay metropolitan region in M. Alam and K.N. Gopi (ed.), *Settlement System of India*, Oxford & IBH, 18-32.

94. Sundaram, K.V. (1967). Settlement pattern for an irrigated colony—An application of the central place theory, *All India Regional Planning Seminar*, University of Mysore.

95. Sundaram, K.V., *et al.* (1972). Spatial planning for a tribal region : A case study of Bastar district, Madhya Pradesh, (Development studies No. 4, Inst. of Development Studies, University of Mysore).

96. Swaminathan, E. (1974). Coimbatore metropolitan area : A study of the hierarchy of settlements, *I.G.J.*, 49 (2), 78-84.

97. Swaminathan, E., *et al.* (1979). Small towns of Coimbatore and Thanjavur districts : A comparative study of patterns, characteristics and hierarchy, in R.L. Singh and R.P.B. Singh (ed.) *Place of Small Towns in India*, NGSI, Regional Pub. 21, 127-42.

98. Thaha, M. (1979). Grouping of settlements, identification of units for block planning and location of socio-economic activities, in *Rural Development in India—Some Facets*, NIRD, Hyderabad, 174-87.

99. Thaha, A. and Thaha, M. (1979). *A Manual on Integrated Rural Development*, (Faizy, Quilon).

100. Thakur, R.N. (1974). Rural service centres in North Bihar, *Ind. Geog. Studies*, 2, 23-27.

101. Varma, L.N. (1973-75). Spatial hierarchy of urban settlements, *Geog. Outlook*, 10, 77-82.

102. Verma, R.S.P., *et al.* (1981). Hierarchy of central places : A case study of Madhubani, *Ind. Geog., Studies*, 16, 53-60.

103. Vishwanath, M.S. (1972). Growth pattern and hierarchy of urban centres in Mysore, *I.G.J.*, 47 (1 & 2), 1-11.

104. Wanmali, S. (1970). *Regional Planning for Social Facilities : An Examination of Central Place Concepts and Their Application*, NICD, Hyderabad.

105. Wanmali, S. (1971). Ranking of settlements : A suggestion, *Behavioural Sciences and Community Development*, 5 (2), 97-111.

106. Wanmali, S. (1972). Review of research studies in India and abroad in L.K. Sen (ed.), *Readings in Micro-Level Planning and*

Rural Growth Centres, NICD, Hyderabad, 235-57.

107. Wanmali, S. (1972). Location of existing and potential growth centres, in L.K, Sen (ed.) *Readings in Micro-Level Planning and Rural Growth Centtes*, NICD, Hyderabad, 196-204.

108. Wanmali, S. (1972). Central places and their tributary population: Some observations, *Journal of Behavioural Sciences and Community Development*, 6 (1), 11-39.

109. Wanmali, S. and Khan, W. (1972). Conceptual and methodological dimensions of central place theory, in L.K. Sen (ed.), *Readings in Micro-Level Planning and Rural Growth Centres* NICD, Hyderabad, 177-88.

12

LOWER LEVEL SERVICE AREAS AND THE HIERARCHY
A Heuristic Appraisal

GOPAL S. KULKARNI AND VINCENT P. MILLER, JR.

THERE has been a pr( )lem of analysis evident in geographical literature for long time. It is one that parallels problems encountered oy those interested in economics and development. It conc-.ns the urban hierarchy. Do we have a full understanding of the nature of the urban hierarchy at its lower levels ? Are we forcing our images on a segment of reality where the hierarchy is either quantitatively or qualitatively different from what we have learned to expect ? Do our methematical and quasi-mathematical concepts so closely related to complementarity and comparative advantage misserve us in the rural environs of the developed world, let alone the underdeveloped world ?

The *New York Times* ran an article entitled, "Shift in Economic Order is the Focus at Cancun." That piece, along with a *Times* editorial, "What's Possible at Cancun", in many ways underscores the aforementioned problem of analysis.[1] A shift in the economic orders calls for a shift in economic thinking, and the possibilities of rural and Third World development are limited by our economic thinking. The call for a new economic order, that was the theme of the Cancun, Mexico World Economic Conference during October 1981, can be interpreted as the climax (hopefully until that time) of the effort to see subsistence economies and their supplementary economic and spatial relationships in a different light. What is possible would seem to depend just as much on our perspective as on the 'economic reality' of hunger, poverty, and exploitation in the First and Third Worlds.

In the field of central place studies and geography in general, the debate over the nature of 'reality' surfaced at the 1960 International Geographical Union Conference held in Stockholm, Sweden. There, the debate took the form of a critique of settlement geography; that is, "does settlement geography have a theory ?" It was the central place theorists who posed the question. Thus, in typical fashion, important empirical contradictions were obfuscated as methodological quibbling in the field. In retrospect, it seems as if the geographers were blaming themselves for the lack of theoretical and empirical correspondence between the conclusions of settlement geography and central place geography.[2]

The 1960s was a time when work began to be published on the spatial and economic nature of the lower level of the hierarchy, both in developed and underdeveloped nations. Some of the work considerably broadened the perspective of the settlement geographers, if not the central place theorists. In this respect, some of the older studies were searched for new perspectives. Even the classic works of Meitzen and Gradmann,[3] long considered somewhat out of style and theoretically irrelevant, were seen to have interesting empirical insights. Thus, Miller in his Ph. D. dissertation, notes that it is possible to trace a continuing development of thought dating from the epical work of von Thünen.[4,5] At the same time, the use of central place theory as a tool of the market forces to generate new market forces (thus, for the haves to exploit the have nots), was proposed and severely criticised.

In the late 1950's and early 1960s, economists, too, were also questioning their own theories of development. One of the important works in this respect was Hirschman's *The Strategy of Economic Development*.[6] From that work, along with those of Myrdal and Prebisch,[7] came the controversy over the means of development, trickle-down or bubble-up ? That is, is it possible to use the hierarchy effectively to spur local development programmes? It would seem that the economists did not allow their debate to become an internecine quarrel, or perhaps that because the economists were read more widely than the geographers, the debate was broadened and so it was not allowed to become a methodological battle. The one book, perhaps more than any other, that took hold of the issue and put it before the world was that of the British planner E.F. Schumacher. That book is *Small is Beautiful*. Slow in its acceptance, but nevertheless remarkably successful, *Small is Beautiful* has prompted a move towards appropriate technology; a type of technology apparently favoured today by the US AID programme in some careful studies of development in South America supported by the agency.[8]

With the general acceptance of appropriate technology and so integrated rural development, the bubble-up mentality became, suddenly, state-of-the-art. The Agency for International Development's *Horizon's* magazine is very careful to note that its programmes are 'state-of-the-art' in these respects. Geographers who were complaining of the use of central place theory, as a tool of capitalism, suddenly found themselves not so theoretically disruptable as they were in the 1960s. Certainly, the publication of Santos' *The Shared Space* (subtitled two circuits of the urban economy in underdeveloped countries), picks up and legitimises McGee's[9] "In Praise of Tradition : Towards a Geography of Anti-Development" though published in English some five years after McGee's essay. Clearly, we, as geographers, have gone beyond a sentimentalist's approach to the preservation of tradition for the sake of tradition. As Santos shows, there are two spatial circuits to Third World urban economies. But what of underdeveloped rural economies? Miller's work in Haiti,[10] alongwith the numerous papers by others on periodic markers, gives evidence of the spatial organisation of the rural underdeveloped realm, if not a solid theory of such development that is equivalent to that of central place theory.

What is possible at Cancun? Certainly our theoretical perspectives have much to say about what is possible. If the "civil religion" of a certain branch of geography is a mathematically oriented hierarchical system, then true to the philosophical insights of Thomas Kuhn, the structure of thought will determine what is possible and feasible.[11] But a revolution in thought has occurred in and out of geography in the last 20 to 25 years, a revolution that is equivalent to the quantitative observations about lower level spatial organisation made by the authors. These observations are the basis for on-going conceptualisation, reflections on research done individually and together in a continuing effort to move towards a theory of spatial organisation for the various niches of the hierarchy. Thus, if we are to have a complete *spatial* theory for a central place hierarchy, then we need to be able to state the nature of spatial organisation at level 1, 2, 3, ...n., of course ours, and the observation of others, do not allow us to do that now for some very good reasons; reasons not only merely related to the paucity of observations per se, but also related to the breadth of our expectation from that theory. The brief review of the developments of the last 20 to 25 years shows that the breadth and scope of analysis has changed. But in the future, if we are to continue to change, not in a willy-nilly way, not depending upon intellectual excitement from outside the field of intellectual faddism, but rather from insights from within the field of geography; how shall we go? The remainder of this essay is a heuristic appraisal stemming from our approach.

TOWARDS FURTHER UNDERSTANDING OF THE CRISIS IN THINKING

What can be said about the crisis in thinking over the means of development, trickle-down or bubble-up, and the implied reliance on or rejection of the theory of the urban hierarchy? One very poignant observation can be brought to the fore here. Traditional hierarchical theory can be said to be 'some what' circular in its thinking, and so 'fallacious'. Thus, certain algorithms dictate what reality should be like; observations from the First World confirm that indeed reality is 'somewhat' like that suggested by the mathematical-geometrical schemas, and so all development should follow, aspire to, the geometry (the morphology) apparently found in First World developed situations. Fallacies in thinking can be hypothesized in two respects of these ideas of development: (i) the empirical evidence seemingly overwhelms the logic of the theory, making the theory a *fait accompli* when we should treat such theory as more heuristic than factual. Because of our empirical success with the theory we have not further examined the logic or the illogic of that theory. Actually, this type of approach to the theoretical problems of central place and trickle-down development mirrors a very, very old debate found throughout science, which is : can all scientific theory be reduced to geometry? Who's geometry should be used, Euclidian geometry or those of others that see more or less than 180° to a triangle? This last comment ushers in a second objection to central place/ trickle-down approaches. That is that (ii) the theoretical-cum-geometrical approach is based upon a particular type of logic. This logic is fundamentally Greek categorical logic. Are we guilty of ignoring the extensive analysis and discussion of the logic of science as developed in the last 20 or 30 years by Braithwaite and many others.[12] In general, the questions can be framed like this: is the world Euclidian? Most especially, is the Third World or the rural environs of the First World Euclidian? Have our geometries-in-mind infact dictated our logic, that is, become a substitute for further analysis of the logical basis of our empiricism?

There are many reasons to think that such is the case. The reason for this does not only reside in esoteric arguments about whether the world is Euclidian. One may hypothesize that it has much to do with quasi-Euclidian notions on the nature of economic systems. For example, a dilemma must be faced in regional planning: shall we plan in the most cost effective or the most spatially equitable way? The *possible* really is determined by the *probable*. This problem becomes acute in the deeply rural First World or throughout the Third World. In the urban developed World, such a problem is hardly noticed, however. This dilemma then, or the avoidance of it, tends to determine the level at which we plan. In general, we plan at that level where the trade-offs of planning, the profits and losses, are favourable. The probability of success must be present or else planning will not proceed. Fiscal and so bureaucratic accountability demands this solution to the dilemma of cost effectiveness versus spatial equality.

But what if we misunderstand the nature of economic system? That is what if our ideas of the probable are simply wrong for part of our economic system, say the deeply rural in the first World, yet correct for other parts of the system. Certainly, this very well might be the case for those areas that have mixed economies such as suggested by the two circuit hypothesis of Santos or the older ideas of circular causation of Myrdal. *Thus, may we say that a theory of the possible demands that all economic landscapes be considered mixed economies*?

Here we must be careful to avoid geographic stereotypes. We all know of such stereotypes. For example, once in doing field work in Norway, one of the authors of this paper was told that Norway could afford a relict landscape. In the Norwegian sense, such a landscape would be related to the concept of *gamalt* or 'old things'. The investigator was in a sense being told that it is good to have landscapes that are different and *that remain different*. The state should take measures to preserve that difference, to maintain living museums of the *stortiden*, of the time of greatness. The details will vary from state to state. Norway has its Kjlen, the United States its Appalachia, Brazil its North, etc. Indeed, the reasons for the toleration of the *gamaltlandskaper* or living museums will vary. In Norway, it may well be rooted in the concept of cultural relativism as reinforced by the language debate of that country. In Brazil and the United States, the reasons may be more purely economic. Certainly, the decentralisation anarchic movement that has swept the world has dignified the differences between people and hence landscapes even in small countries. These differences may blind us through a type of stereotypic persuasion that goes something like this: "They are different because they are different. Our (economic, social, political) norms are not their norms. Therefore, what is *probable* (and thus *possible*) for us, is not probable for them." It becomes a live-and-let-live attitude with respect to the landscape, in part based upon the acknowledgement of differences, in part based upon stereotypes, or so it may be hypothesized.

Such stereotypes, however are not to be the basis of theory because they certainly are not value free. While it is debateable that any theory can ever be value free, we must strive to recognise the value limitations of our hypotheses. Static theory of the kind described above suffers from many value limitations, as the more radical thought is quick to remind us. We may never be able to shed the problem of stereotypes, but an approach towards managing the problem is to move to a more dynamic theory. For example, research in south Norway, by one of the co-authors of this essay, presents a case for an ebb and flow of a mixed economy. That is, at times the subsistence economy is much stronger than at other times as if the *gamalt* or old things, the relic landscapes, become defensive and resistant to change, and then at other times willingly submit to change. The stereotype of an archaic society with its ancient language is theoretically misleading. Through an extension of this idea, is the stereotype of an exchange and subsistence economy, or of the upper and lower circuit of the economy, theoretically misleading? Well might it be if the analysis is that of static theory.

In what way is the stereotype of the upper and lower circuit of an exchange and subsistence economy *theoretically* misleading? Clearly, it may be misleading because we approach our landscape laboratories with a pat, unyielding, assumption about the nature of reality and then force that reality into our preconceived moulds using something of a dichotomous key. If not exchange, then subsistence; if not upper circuit, then lower circuit. A dynamic theory of settlement morphology would free us from our static geometries and unyielding assumptions, but not remove all of stereotypes that plague us. Moreover, it would be rewarding in that we can allow the theoretical inconsistencies seen in the landscape to feed back as a corrective to theory. It will allow us to revise our hypotheses and then to vary our planning paradigms to fit the particular sector that is to be developed.

But, and a very important proviso indeed, does not such

dynamic theory necessarily involve historically oriented studies, studies that are difficult or impossible in underdeveloped realms? The authors see that there are two approaches to this problem. First, geographers could do regional development monographs perhaps very similar to the famed *Geographie Universelle*, but with a much sharper focus. Thus, a historically oriented development monograph would trace the historical geography of changes in accessibility, or the historical geography of landuse, pinpointing critical junctures in landuse change in this way. One of the co-authors of this paper has attempted such an approach for both an area of south Norway and Haiti and has found the results rewarding, even in an area of such poor documentation as Haiti. In short, such studies are not impossible even in the most underdeveloped state in the Americas and are certainly feasible in many other areas. This approach is seen as fully equivalent to the valuable ethnographic monographs produced on numerous cultural realms by historically oriented anthropologists.

The second approach is to uncover and catalogue the 'pathology' of development malaise. In this sense, the geographer would look for evidence of varying, even contravening, structures in the landscape, much as Santos has done in his *The Shared Space*. The geographer would become a type of sleuth looking for tale-tale signs of unresolved change and then prescribe a treatment that would minimise the effects of change (so that the medicine is not worse than the illness) and maximise the rewards of change (so that the illness of underdevelopment is treated). Like any doctor attempting to bring about a cure, we must be wary that we aren't treating a normal situation and so must not mistake pathology for normalcy. This requires a knowledge particularly of the flows of subsistence economy. Here, then, we need to understand the upper and lower limits of the normal commerce, the limitations placed on the society by accessibility, the historical significance of the landuse patterns and traditions, and so forth.

CLUES TO THE NATURE OF A DYNAMIC SPACE ECONOMY IN A RURAL FIRST WORLD SETTING

With this background and orientation in mind, the authors sought to examine the lower level of the hierarchy in a First World setting. The area is three counties in Western Pennsylvania. Western Pennsylvania is an older industrialised region that has been famous for its coal deposit and heavy industry concentrations, particularly around Pittsburgh. The three county region studied is some forty to sixty miles north-east of the Pittsburgh conurbation.[13] The region is more or less rural and in places economically depressed ; in others economically healthy. It is an interesting region because of the marked contrasts found throughout places of wealth and poverty in the urban shadow of Pittsburgh.

In a paper published in 1980, the authors presented a case for using central place service areas as planning units rather than counties or MCDs (minor civil divisions, such as boroughs, townships, or other political and civil administrative units). This point of view comes from the very obvious conclusion that if spatially equitable planning is to be aspired to, then service areas (regardless of whether they overlap civil administrative units) should be on a basis for planning. In a previous study care was taken to map the structure of the major central place service areas of the three counties.[14] The approach used in these studies was both that of field work and also of collecting data from store change accounts, newspaper circulation and so forth. This data was then placed on a base map of MCDs and analysed. Various analytical techniques were used including Davies' functional index [15] and cluster analysis. The cluster analysis and the calculation of the functional index are two different ways of utilising the same data. This data is a list of the array of services available at the various central places of the region (given in *Appendix I*).

Table 12.1 shows the functional index for the central places of the three county area. The demarcation "city", "town", etc., was determined qualitatively based on rather obvious breaks in the data. Somewhat later in point of time,

TABLE 12.1 : **Functional index of central places in Butler, Armstrong, and Indiana counties**

	Butler County		Armstrong County		Indiana County	
	Central Place	Functional Index	Central Place	Functional Index	Central Place	Functional Index
1	2	3	4	5	6	7
City	Butler	5435.68	Kittanning	4135.91	Indiana	5471.78
Town	Slippery Rock	604.81	Ford City	1088.08	Blairsville	1173.22
			Apollo	820.75		
Village	Evans City	340.70	Freeport	457.17	Homer City	506.12
	Chicora	259.90	Leechburg	320.32	Clymer	138.26
	Portersville	174.18	Rural Valley	216.85	Saltsburg	103.10
	Zelienopole	163.64	Worthington	211.51	Cherry Tree	80.53
	Lyndora	153.17	Cowansville	89.40	Marion Center	79.80
	Valencia	154.45	Manorville	71.16	Black Lick	66.60
	Harrisville	136.43	Elderton	61.04	Plumville	59.29
	Mars	128.04	Templeton	56.20		
	West Sunbury	117.17	Yatesboro	53.65		
	Prospect	101.04				
			Distant	42.49		
Hamlet	Karns City	84.66	McGrann	39.34	Creekside	42.46
	Saxonburg	76.63	Nu Mine	37 08	Hillsdale	39.93
	Harmony	70.28	Cadogan	33.75	Glen Campbell	34.24
	Renfrew	57.71	Adrian	25.26	Penn Run	30.43

	Petrolia	52.56	Sagamore	21.95	Clarksburg	23.30
	Eau Claire	49.95	Atwood	13.75	Rochester Mills	22.56
	Meridian	47.94	Kaylor	10.91	Dixonville	21.96
	Boyers	44.14	Craigsville	10.05	Commodore	21.56
	Sarver	35.94	Widnoon	9.09	Shelocta	20.95
	Bruin	26.02	Margaret	6.87	Armagh	20.03
	Cabot	22.38	Schenley	6.06	Robinson	18.34
	Callery	16.95	Parker	5.88	Smicksburg	15.92
	Connoquenessing	13.15	State Lick	5.26	Home	14.34
	Forestville	12.87	Spring Church	5.22	Starford	12.65
	Hilliards	10.33	Ford Cliff	4.29	Ernest	12.50
	Cooperstown	5.61	S. Bethlehem	3.33	Heilwood	12.16
	Branchton	5.47	Brady's Bend	3.07	Coral	11.57
	North Washington	5.23	Riner	3.07	Alverda	11.32
	West Winfield	2.17	South Bend	2.70	Josephine	9.69
			Belknap	1.85	Marchand	9.12
Hamlet					Chambersville	8.75
					Lucern Mines	8.61
					McIntyre	8.53
					West Lebanon	7.73
					Graceton	7.47
					Arcadia	6.90
					Clyde	6.07
					Huff	5.68
					Strongstown	5.50
					Rossiter	5.26

TABLE 12.1 : **Contd.**

1	2	3	4	5	6	7
					Pine Flats	4.76
					Wandin	4.41
					Kent	4.38
					Brush Valley	4.34
					Beyer	3.63
					Gipsy	3.88
					Adams	3.57
					Aultman	3.57
					Tunnelton	2.60
					Larimer	2,05
					Coal Run	1,82
					Brenizer	1.32
					Iselin	1.06
					Cookport	1.01

Source : Authors' Computations. Also see *Appendix I*.

the same data was subjected to cluster analysis. Table 12.2 shows the amount of information lost in going from one group to another in the cluster analysis.

In doing the cluster analysis all the data for all of the towns, villages, etc. for the entire three county area were processed together (they were not grouped together in computing and assigning the hierarchy for the individual counties, see Table 12.1). The first group for the entire study region consisted of the three largest central places which dominated the services picture for the region. In short, the Davies functional index approach was used to understand the

TABLE 12.2 : **Effects of successive settlement groupings**

Groups	Information loss at this stage (%)	Cumulative information loss (%)
1	69.26	100.00
2	8.94	30.74
3	6.29	21.80
4	3.99	15.50
5	2.64	11.50
6	1.46	8.86
7	1.27	7.40
8	1.06	6.13
9	.55	5.07
10	.52	4.52
11	.48	4.00
12	.42	3.52
13	.36	3.10
14	.34	2.74
15	.26	2.40
16	.19	2.14
17	.18	1.95
18	.15	1.77
19	.14	1.62
20	.13	1.48
21	.12	1.35
22	.11	1.23
23	.085	1.12
24	.071	1.04
25	.067	.97

Source : Cluster analysis.

hierarchy of a particular county and the cluster analysis was used to give conclusions about the entire study region.

Both approaches have their place and their value. The delimitation of the functional index is practical from the point of view that development funds are allocated at the county level and then redistributed to minor civil divisions (MCDs) as need determined by county officials. The cluster analysis treats the whole area as one unit which is practical in terms of the known spatial interaction between its central places. The "picture of reality" of each then comes from a different scale of analysis, one (the cluster analysis) is considered to be more holistic, the other more anarchic or autarkic (the functional index for the individual counties).

Accordingly, planning scenarios could differ greatly depending upon the scale used in developing a plan. No studies of individual county planning practices have been attempted at this stage of the research. However, it is not difficult to conjure up a scenario made from the perspective of the county as the planning unit as compared to that of the three county area as the planning unit. County planning directors might well become sensitive to local political pressure and so fortify the higher order central places with so-called missing vital services. The county directors then may be said to respond to cost-effective planning approaches. A director for the three county area may be more concerned with the lower end of the hierarchy so as to provide a minimal service base for the rural poor. This approach would presumably identify the poorly served areas in the three county region and then take steps to change these situations.

It might well appear that the two approaches (the holistic and the anarchic/autarkic) represent radically different means of planning, just as capitalistic and Marxist geographers are radically different in their approaches. From the point of view advanced here, such differences in approach are merely differences in degree but not in kind. Both have as their basis for action a particular but different static view of the space economy. Their difference is a scale difference rather than a true philosophical difference.

How may we plan effectively for such a region without succumbing to "degree-differences" (see the above paragraph) in approach? We must search for evidence of the dynamics of the space economy. In doing this, we must continue to use the functional index and cluster analysis techniques. We must develop yardsticks from these and other techniques that point to symptoms of malaise in the space economy. The holistic approach may well be no better than the anarchic/autarkic approach if a viable subsistence economic system is subverted in the name of planning or in the name of a pet pseudo-philosophy.

What are some of these yardsticks? The answer to this is probably regionally specific rather than theoretical at this point in the development of our ideas. Different regions will have different yardsticks. Work in Norway indicated that a *semi-functional* landuse tradition, that of the seter system, was (and to some extent is) used as a means of preserving the subsistence economy[19]. The point is, we must have the wisdom to discern *semi-functions*. *Semi-functions* are those functions that are used discontinuously through time as a means of preserving regional autonomy and fending off perceived undesirable centralising forces. We must also learn to recognise the functional attributes of supplementarity and a subsistence economy.

In the three county study area the authors began to recognise that forces emanating from the Pittsburgh region affected different parts of the study region in different ways even though those areas are similar in terms of the functional index. This is a 'feeling' garnered from field work rather than a conclusive piece of evidence seen from the data itself. There is a need to do such field work and then go back to the data and look for tale-tale signs of say, resistance to change or for a type of self-sufficiency that enables "one to take care of one's own". For example, very recent analysis indicates that there is very great resistance upon the part of some rural poor (the so-called "new poor" in the area) to welfare type programmes. In plain truth, certain people simply refuse to apply for welfare. This attitude may or may not be economic suicide, depending upon the presence of certain semi-functional services. In other cases, the authors have determined that there are some 'residual' functions such as frozen food lockers (used by those who don't have freezers or refrigerators) or tourist homes (for inexpensive overnight lodging) that are to be found in the poorer reaches of the study area.

Throughout the region, a very considerable underground economy has grown up. The evidence of this is great if only we learn to recognise it. For example, many people grow much of their own food in small backyard gardens, buy clothes and housewares in a myriad of sales held in garages, living rooms and lawns (variously called garage sales, house sales, lawn sales, etc.). Families are joining together to share lodging costs (families of sons and daughters are moving in with parents, reconstituting the extended family tradition), sharing transportation costs through car pools and so forth. This underground economy (so-called because the financial transactions most often escape the eye of the tax man and so sales taxes, income taxes are not paid) is to be seen as the developed equivalent of Santos' lower curduit in the under-developed world. In short, the underground economy is evidence of the dynamics of the rural economy in the study area. Would the laws of supplementarity or complimentarity then apply? This sort of economic activity is a buffer for hard times, (vis-a-vis seters in Norway), which this region has experienced in the last ten or so years. It would seem that the laws of supplementarity rather than complimentarity should apply. The hypothesis presented here is that once times become better the garage sales and other attendant activities will decline and new items rather than used, separate housing rather than joint occupancy and so on, will become the norm.

SUMMARY AND CONCLUSIONS

Towards a recognition of the dimensions of the lower levels of the hierarchy

The paper is a plea for more thoughtful analysis and planning at the lower levels of the urban hierarchy. Also, the suggestion is made in a very pointed way that the rural areas in the developed world are not so very different from those of the underdeveloped world. If there is a great difference it is a difference of degree and not of kind.

Our analysis must be cognisant of the dimensions of development. We may easily misunderstand the development that has already occurred because of our preconceived notions of what development and progress is all about. One very important dimension to be considered in the analysis of the lower levels of the urban hierarchy is that of *economic state* of the area studied. Are the forces of complimentarity or supplementarity more prevalent? That is: where is the area with respect to the continuum between exchange and subsistence? That becomes a very important question in our analysis and any subsequent practical application in the form of planning. Planning may well do more harm than good because we fail to recognise the elements of stability in a subsistence situation.

We must also consider the resilience of the area studied. To what degree are the people able to respond to economic distress and to change from outside the region? These dimensions are certainly two sides of the same coin and must be carefully assessed. Subsistence systems as well as exchange systems have a dynamic all their own and we must further our understanding of those dynamics.

We must carefully research the structure of the service areas of the lower level systems by detailing the linkages through space that exist. This topic has been only mentioned here and is indeed a very rich and important realm for further research, a topic worthy of extensive investigation. Ideally, in time, we can come up with models of the different types of service areas and so a typology of the service areas found on the continuum of exchange to subsistence economies.

So far, only the object of analysis has been discussed. We cannot ignore the subject of our analysis. We must decide what we are about. What is the subject that we desire to pursue? The radical geographers are perfectly correct in bringing this question before us. We must ask ourselves, not in the name of a particular humanistic philosophy but in the name of seeking truth and the betterment of mankind, what our goals are to be. The dichotomy presented here is that of the goal of cost efficient planning versus spatially equitable planning. Perhaps these categories can be improved and expanded upon. Regardless of how we perceive our goals we should have the wisdom to understand that, just as you cannot separate subject and object, our goals will affect the reality that we deal with.

REFERENCES

1. Bernard D. Nossiter, Shift in economic order is the focus at Cancun, *New York Times*, 21 October 1981, p. A 1-6; and What's Possible at Cancun, *New York Times*, 22 October, p. A-26.
2. This paragraph is in reference to W.L. Garrison's (1962) comments contained in Knut Norberg ed., *Proceedings of the I.G.U. Symposium in Urban Geography*, Lund C.W.K. Gleerups pp. 463-464.
3. A more complete analysis of this development is to be found in V.P. Miller, Jr. (1970), *The Impact of Isolation on Urban-Rural Relationships in Setesdal, South Norway*, Michigan State University, Summer, available from dissertation abstracts, Ann Arbor, Michigan.
4. *Ibid.*
5. The pros and cons of the points outlined in this paragraph are more thoroughly reviewed in V.P. Miller and G.S. Kulkarni's *The Central Place Hierarchy and Accessibility to Services in Butler, Armstrong and Indiana Counties, Pennsylvania*, 1976, Centre for Community Affairs, Monograph 1, Chapter 1. The opinion given in the parenthesis is a typical critical remark of the use of central place theory as an instrument of planning and does not necessarily represent the views of the authors.
6. A.O. Hirschman (1958), *The Strategy of Economic Development*, New Haven : Yale University Press.
7. Gunnar Myrdal, 1965, *Challenge to Affluence*, New York : Vintage, and Raul Prebisch, (1964), *Towards a New Trade Policy for Development*, New York : United Nations.
8. For example the "Lessons Learned" section of the Agency for International Development's *Horizons* magazine reviews state of the art methods in the area of development.
9. Milton Santos, (1979), *The Shared Space*, New York : Methuen, and T.G. McGee, (1974), In Praise of Tradition : Towards a Geography of Anti-Development, *Antipode*, 6-3 (Dec, pp. 30-49).
10. V.P. Miller Jr., (1982), *Project Ebenezer : Modelling Holistic*

Missions, Ministries In Action, P.O. Box 340325, Coral Gables, Florida 33134. This is a monograph assessing the problems of Christian development missions in the country of Haiti, W.I.

11. Thomas Kuhn (1970), *The Structure of Scientific Revolutions*, Chicago : The University of Chicago Press.
12. R.B. Braithwaite (1960), *Scientific Explanation*, New York : Harper and Brothers, 1960; and Kuhn, op. cit.
13. Vincent P. Miller Jr. and Gopal S. Kulkarni, (1979), 'A case for central place service areas as planning units.' Giles and Hershey, eds., *Shaping the 80's*, Proceedings of the Pennsylvania Community Development *Society*.
14. op. cit., fnt. 5.
15. Wayne K.D. Davies, Centrality and the central place hierarchy. *Urban Studies*, 4, pp, 61-79.
16. The seter system is transhumance in Scandinavia. Analysis indicates that the use and disuse of this system through time correlates well with up- and down-turns of the economy, that is when times are good the seters fall into disuse, when they are bad, the seters are used. Thus the seters are roughly one of the Norwegian equivalents of the underground economy found in America. See Miller Jr., (op. cit., fn. 3) Chapter IV "The Geography of Exchange and Subsistence in Sorlandet and Setesdal."

APPENDIX-I

LIST OF VARIABLES

Sl. No.	Variable	Sl. No.	Variable
1.	Advertising	44.	Junk, salvage, and wrecking
2.	Agricultural supply and machinery	45.	Lawyer
3.	Aircraft and service	46.	Lumber supply
4.	Appliances, supply and service	47.	Mail order
5.	Auto dealer, New	48.	Medical laboratory (independent)
6.	Auto dealer, Old	49.	Medical equipment
7.	Auto repair	50.	Mobile home
8.	Auto supply	51.	Musical instruments
9.	Baker	52.	Newspaper and publishing
10.	Bank	53.	Nurses and medics
11.	Bar. tavern and beer sales	54.	Nursery and landscaping
12.	Barber and scalp treatment	55.	Nursing home
13.	Beauty saloon	56.	Office equipment
14.	Bookstore	57.	Optometrist
15.	Building supplies	58.	Pets, care and supply
16.	Burial and funeral director	59.	Photographer
17.	Chiropractor	60.	Post Office
18.	Church	61.	Printing
19.	Cleaning and laundry service	62.	Professionals (other)
20.	Clergymen*	63.	Public library
21.	Clothing	64.	Radio dealer
22.	Communications	65.	Real estate agent
23.	Credit and loans	66.	Recreation
24.	Dairy and poultry	67.	Restaurant
25.	Dentist	68.	Retailing (other)
26.	Department store	69.	Rubbish and garbage removal
27.	Physician and surgeon	70.	Service station
28.	Education	71.	Shoe store
29.	Energy supply	72.	Social organization
30.	Florist	73.	Sporting goods
31.	Frozen food locker	74.	System maintenance
32.	Furniture	75.	Tailor
33.	General contractor	76.	Telephone answering service
34.	Gift and hobby	77.	Television sales and service
35.	Government	78.	Tourist home
36.	Grocery	79.	Trading stamp company
37.	Hardware	80.	Transportation
38.	Hospital	81.	Travel agency
39.	Hotel and motel	82.	Variety discount store
40.	Industrial equipment	83.	Veterinarian
41.	Insurance adjustor	84.	Water supply company
42.	Interior decorator	85.	Wholesale (supply and service)
43.	Jewelry and watches		

*Not used in this study as data was incomplete.

13

SERVICE PROVISION AND SERVICE CENTRES
Lessons from Miryalguda*

SUDHIR WANMALI

THE ANALYSIS of geographical space and its interaction with other spaces in India in general, and in Andhra Pradesh in particular, has never been far from the centre of attention, and scrutiny, of Professor Prakasa Rao. His numerous scholarly publications on this topic have always been a source of encouragement to many young geographers (Rao, 1961, 1963, 1964, 1964a, 1966, 1969, 1974, 1975, 1975a, 1979) ; the author of this article being one of them. Although the piece which follows is unlikely to match the high standards of Professor Rao's contributions, it is submitted as a tribute to him from one who has had the good fortune to be professionally associated with him during the last decade and a half.

The article is placed within the growing field of interest in government policy and its impact on the provision of services in rural India. It is based on the empirical evidence from Miryalguda Taluka, of Nalgonda district, Andhra Pradesh.

*The article is abstracted from a report on "Service Provision and Rural Development in India : Miryalguda Revisited," prepared by the author and published by the International Food Policy Research Institute, Washington, D.C. The views expressed in this article are those of the author and do not necessarily reflect those of the Institute to which he belongs.

The taluka is situated in the command area of Nagarjunasagar Hydro Electric Project (hereinafter referred to as NSP) on the River Krishna. A part of the study area cannot be irrigated and is, therefore, dry (Fig. 13.1). The analysis of the development of service provision in the irrigated and the dry tracts of the taluka can be of some interest to the administrators, development planners and academics. What

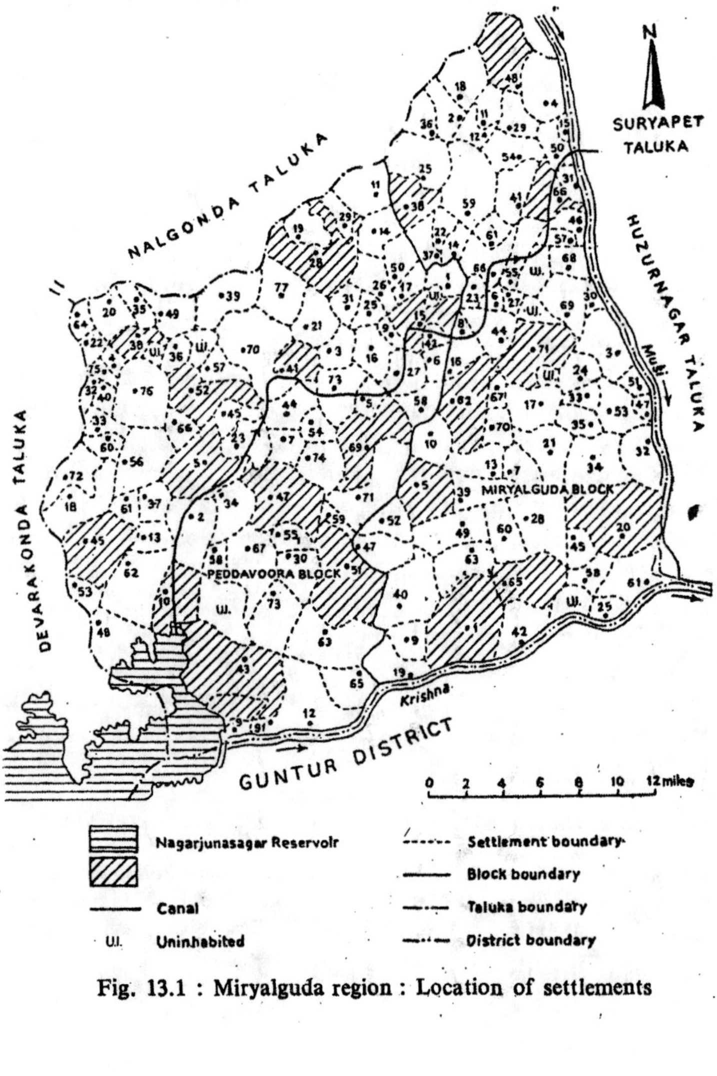

Fig. 13.1 : Miryalguda region : Location of settlements

makes the discussion interesting is that it studies change in the same region ; and in the same villages ; after a gap of ten years. In this entire enquiry, an emphasis is placed on the study of underlying relationships between the concerned aspects in their static (in 1968 and in 1978) as well as their dynamic (between 1968 and 1978 as a time period) senses.

Miryalguda Taluka was a drought prone area which was provided with irrigation waters. The cropping pattern of the study area has undergone a structural change as a consequence of introduction of these waters. This change involves extension, intensification and diversification of cropping which was facilitated by simultaneous provision of credit, banking, transport, marketing, storage, communications, animal husbandry and input distribution services as well as those in education and health. The employment potential generated, and higher incomes earned, were responsible for the demand also of various retail services such as kirana stores, general provisions, cloth, hardware, fertilisers and pesticides, restaurants, chemists and markets to sell agricultural produce (and for buying of urban consumer goods).

In the following paragraphs an attempt is made to examine the patterns of service provision with respect to locations of rural (government sector) and retail (private sector) services, their use, service centres, service areas and service populations in the study area. An attempt is also made to note changes in the pattern and compare them in the context of recommendations made by an earlier study (Sen and others, 1971) ; particularly in the areas of demographic, functional and spatial features of the service centres.

SERVICE PROVISION

Services and their operational definitions

There are twenty-nine services in the taluka which can be designated as central functions. This is because these services are 'available in a few places and are availed of by a number of places' (Christaller, 1966). For the sake of the present exercise let education, health, credit, banking, transport, communications, animal husbandry, trade and marketing and distribution of agricultural inputs be defined as rural services (whose provision is controlled/facilitated by the government). Similarly, let retail services of various kinds be a separate group (whose provision is controlled/facilitated by the private sector). This categorisation is not mutually exclusive. For example, there are some services in the former category, such as health, transport, and distribution of fertilisers, pesticides and agricultural machinery, where the private sector operates alongside the government sector. Similarly, the trading of agricultural produce which was until recently a spontaneously organised system of traders, consumers and weekly markets is now being complemented by a system of government controlled regulated markets.

The reasons why government alone appears to be the main provider of these services, at least in the initial stages, are many but the following can be mentioned by way of examples. In certain services (such as education and health), in the short run, the private cost is higher than the private gain whereas in the long run, the social cost is lower than the social gain. In others (such as supply of agricultural inputs, animal husbandry, transport facilities, credit and banking), there are some initial risks which the private sector tends to overestimate ; and this requires the government to enter the business of their provision. In some other cases (such as marketing, roads, processing units) the externalities of the infrastructure for creation/provision of these services cannot be captured by the private sector and hence the government looks after their provision too.

Once the government has played this pioneering and facilitating role it is quite possible, as will be seen later, that this role is then complemented by the efforts from the private sector also.

Some of the services such as general provision stores, chemists, retail cloth stores, hardware shops, restaurants, posts and telegraph offices, primary, middle and secondary (high) schools, tea and coffee shops and weekly markets hardly require any further clarification of what these mean. Others need some clarification regarding their scope and that is given below:

A *branch post office* sells stamps, envelopes, postcards, inland letters and airmail letters ; receives and sends money orders and has postal savings and banking facilities. A *sub-post office*, in addition to the above, sells postal orders, receives and sends telegrams and issues radio licenses. A *primary credit society* arranges for agricultural loans for its members, who belong to specified villages. A *co-operative bank* has the jurisdiction for providing agricultural finance to the residents of the taluka ; it normally provides this through the primary credit societies but sometimes it also makes the loans available directly ; A *rural bank* looks after agriculture related finance but not necessarily for agricultural production. A *nationalised bank* (if it is a branch of the lead bank for the district) looks after the entire financial need of the taluka. Although the *animal husbandry centre* is meant to serve only those villages which are within its jurisdiction, sometimes animals from villages outside its jurisdiction are also treated. *Seeds distribution* is controlled solely by the panchayat samiti for all the villages which are located within it ; but *fertilisers, pesticides* and *agricultural machinery* are distributed by both the government and private sectors. The *regulated market* and its sub-yards are meant for purchasing of paddy, castor, pulses and groundnuts. Other products are not regulated and these are traded in the weekly markets. *Retail kirana stores* sell daily needs at the village level.

Measure of inter-service and intra-service importance

Although twenty-nine services are performed by the settlements in the taluka as a whole, this does not mean that all settlements perform them at an identical level of complexity. For example, education as a service exists in almost all villages ; but within that primary schools are more ubiquitous and there is only one junior college in the taluka. While measuring the importance of a settlement in education service, if the settlement has a secondary school which also incorporates a middle and a primary school in the same building, then the secondary school alone was considered. However, if in addition to the secondary school, there also exists a middle or a primary school, then these were considered separately. The bus service was weighted by the frequency of its operation.

An attempt was made to note the population size at which a service first occurred in the settlement system of the taluka. This point of entry of a service, in terms of population size, has been used in a number of other studies; and has been defined as the threshold of the service. Table 13.1

TABLE 13.1 : **Rural and retail services : thresholds of population**

Services	Thresholds of population
1. Primary school (government only)	250
2. Retail kirana stores (private only)	260
3. Primary credit society (government only)	260
4. Centre for medical checkup (government only)	290
5. Bus stop/service (private and government)	320
6. Retail cloth stores (private only)	340
7. Branch post office (government only)	550
8. Tea and coffee shop (private only)	620
9. Fertiliser and pesticides shop (private only)	960
10. Middle school (government only)	1,390
11. Hardware shop (private only)	1,390
12. General provision shop (private only)	1,390
13. Sub-post office (government only)	1,390
14. Animal husbandry centre (government only)	1,400
15. Secondary/high school (government only)	2,060
16. Allopathic treatment/clinics (private and government)	2,110
17. Restaurant (private only)	3,090
18. Weekly market (private only)	3,700
19. Rural bank (government only)	4,400
20. Chemists (private only)	4,500
21. Junior college (government only)	16,151
22. Surgery and hospitals (private and government)	16,151
23. Posts and telegraphs (government only)	16,151
24. Nationalised bank (government only)	16,151
25. Cooperative bank (government only)	16,151
26. Seeds distribution centre	16,151
27. Fertilisers and pesticides distribution centre (government only)	16,151
28. Veterinary hospital (government only)	16,151
29. Regulated market (government only)	16,151

Source : Wanmali (1971), Census of India, Nalgonda District, 1971 and Field Survey, 1978-79.

gives the population thresholds of the services in Miryalguda taluka. An attempt was then made to ascertain the relative importance of services based on the threshold values. This was done in the following manner : the threshold value of the lowest order service was considered as a base with a value of 1. Inter-service importance was obtained by calculating the values of other services from this base. For example, the value of primary school was 1 ; that for the branch post office was 2.20. This is because the threshold population of the primary school was 250 and that of the branch post office was 550 (Table 13.2). The intra-service importance, between different elements of the same service, was also obtained in a similar fashion. The middle school with a threshold value of 1,390 was 5.56 times more important than a primary school (Table 13.3).

TABLE 13.2 : **Inter-service importance : Weights of services**

Services	Population	
	Threshold	Weights
Education		
Primary school	250	1.00
Credit/Banking		
Co-operative society	260	1.04
Health/Medicine		
Centre for medical checkup	290	1.16
Transport		
Bus stop	320	1.28
Communications		
Branch post office	550	2.20
Agricultural Inputs		
Fertiliser shop	960	3.84
Retail Shops/Private		
Kirana stores	260	1.04
Cloth stores	340	1.36
Tea and coffee shop	620	2.18

Source : Field Survey and Census of India, Nalgonda District, 1971.

TABLE 13.3 : **Ranking of services**

Service	Population Threshold	Population Weight
1. *Education*		
Primary school	250	1.00
Middle school	1,390	5.56
High school	2,060	6.40
Junior college	16,151	64.60
2. *Credit/Banking*		
Primary credit society	260	1.04
Rural bank	4 400	17.60
Co-operative bank	16,151	64.60
Nationalised bank	16,151	64.60
3. *Health/Medicine*		
Centre for medical checkup	290	1.16
Allopathic treatment/clinic	2,110	8.44
Surgery and hospitals	16,151	64.60
4. *Transport*		
Bus stop/service	320	1.28
5. *Communications*		
Branch post office	550	2.20
Sub-post office	1,390	5.56
Post and telegraph office	16,151	64.60
6. *Agricultural Inputs*		
Fertiliser and pesticide shop	1,470	5.88
Seeds distribution centre	16,151	64.60
Fertiliser and pesticide distribution centre	16,151	64.60
7. *Marketing of Agricultural Produce*		
Weekly markets	3,700	14.80
Regulated markets	16,151	64.50
8. *Veterinary Services*		
Animal husbandry centre	1,400	5.60
Veterinary hospital	16,151	64.60
9. *Retail Goods*		
Retail kirana stores	260	1.04
Retail cloth stores	340	1.36
Tea and coffee shop	620	2.48
General provisions	1,390	5.56
Hardware	1,390	5.56
Restaurant	3,090	12.60
Chemist	4,500	18.00

The values of inter- and intra-service importance were used as weights to measure the value of total service structure at a particular settlement. Another variable, the number of such services, was added to the scheme of ranking at this stage. If a settlement had one primary school then it was given a value of 1 ; if it had ten schools, its value was 10.

It is felt that by considering the variety, level and number of services performed by all settlements it is possible to arrive at the measure of importance of settlements in the taluka.

Changes in service provision

It was seen in the study area that the provision of irrigation facilities has resulted in both quantitative and qualitative changes in the agriculture of the taluka which involve extension, intensification and diversification of the cropping pattern. This change in agriculture was responsible for a greater demand for both rural (government controlled) and retail (provided by the private sector) services in the taluka. This demand in its turn was responsible for quantitative, and qualitative changes in the provision of services in the taluka.

Regionally, there has been an impressive increase in the spread of rural and retail service provision. Table 13.4 lists the number of settlements which have these services. The number of times a particular service is available in a settlement is, however, not noted in this table. The notable feature of this table is the increase in the number of settlements providing the lower and middle order services during the years 1968-78. In addition, some new services, or channels for their provision, have also been widely encountered; these include rural banks and regulated markets (and their sub-vards) in the government sector and allopathic clinics, buses and trucks and fertiliser and pesticides shops in the private sector (in addition to those which are provided by the government).

There were 23 settlements in the taluka in 1968 which did not have any service provision at all; this number is down to 12 in 1978. The number of settlements which did not

TABLE 13.4 : **Occurrence of rural and retail services : 1968-78**

Name of the service	Number of settlements with service	
	1968	1978
Primary school	111	128
Primary credit society	79	107
Retail kirana stores	46	116
Branch post office	46	61
Bus stop/service	42	52
Centre for medical checkup	31	32
Tea and coffee shop	24	54
Fertiliser and pesticides shop	16	33
Allopathic treatment/clinics	12	16
Secondary/high school	10	23
Middle school	9	4
Retail cloth stores	9	40
Animal husbandry centre	6	24
General provision stores	5	14
Sub-post office	4	6
Chemists	2	5
Allopathic treatment and clinics	3	21
Hardware shop	2	5
Restaurant	2	7
Nationalised bank	2	3
Junior college	2	2
Surgery and hospitals	2	3
Weekly market	1	5
Regulated market	1	3
Veterinary hospital	1	1
Posts and telegraphs	1	1
Seeds distribution centre	1	1
Fertiliser and pesticides distribution centre	1	1
Co-operative bank	1	1
Rural bank	0	6

Source : Sen and others (1971), Field Surveys, 1969 and 1978-79.

have any rural service (government sector) in 1968 was 26, which is down to 17 in 1978. But the most impressive spread was in the retail services (private sector). In 1968, 104 settlements did not have any retail establishments;t his number is down to 34 in 1978. Admittedly, the form of government and private sectors' involvement in the provision of services is basic and on a need to establish basis. For example, the first rural service to appear in the settlement system of the taluka is normally a primary school and the first retail service of a similar nature is normally a kirana store.

A large number of new services were established during the same time period. Thirty new primary schools (15 of these being in settlements which needed an additional school of this type), 2 middle schools, 7 secondary (high) schools, 29 primary credit societies, 8 rural banks, 14 branch post offices, 13 centres of fertiliser distribution, 7 centres of pesticides distribution, 10 animal husbandry centres, and 1 regulated market (with two sub-yards) were established by the government. In the private sector, about 514 kirana stores, 114 retail cloth stores, 78 general provision stores, 12 hardware stores, 21 chemists shops, 30 restaurants and 216 tea and coffee shops were established in the taluka. In addition, three weekly markets also appeared on the rural scene in addition to the two which already existed.

The average value of change in the total service provision of the taluka, for the period 1968-78, was 30.61; the average values for changes in the rural and retail services were 15.83 and 14.78, respectively. In the taluka, there are 25 settlements which have values above the regional average; these are : Adividevulapalle, Alagadapa, Amangal, Anjanpalle, Anumula, Damercherla, Kondrapole, Miryalguda, Madgulapalle, Nidmanoor, Peddavoora, Perur, Devulapalle, Thungathurti, Tadakamalla, Tungapahad, Tirumalgiri, Tripuravaram, Thummadam, Vadepalle, Veerlapalem, Vemulapalle, Vijayapuri, Yacharam and Yerraballi. Of these, all but seven are in the irrigated tract. Only one among the exceptions has an advantage of benefiting from a lift irrigation scheme.

The correlations between changes in irrigated land, cropped area, population, total services, rural services and retail services were calculated (Table 13.5). It is very interesting to note that the first two and the last four are statistically highly correlated. The immediate impact of the

TABLE 13.5 : **Correlation coefficients between changes in irrigation, cropping, population and services**

Irrigated area	1.00*	0.50*	−0.01	−0.03	−0.01	−0.02
Cropped area	0.50*	1.00*	−0.02	−0.03	0.01	−0.02
Population	−0.01	−0.02	1.00*	0.88*	0.86*	0.88*
Rural services	−0.03	−0.03	0.88*	1.00*	0.96*	0.99*
Retail services	−0.01	−0.01	0.86*	0.96*	1.00*	0.99*
Total services	−0.02	−0.02	0.88*	0.99*	0.99*	1.00*

*Statistically significant at 0.01 level.

changes in irrigated facilities (that is providing irrigation to the area) was in the field of cropping pattern which changed from dry to irrigated farming where more numbers of crops were grown, some of the changes in cropping were facilitated by the provision of services in both the government and the private sectors. This increase in production created employment opportunities (on a permanent as well as seasonal base) and raised incomes in the region. More people have settled down in the irrigated tract of the study area. Since the services were located in settlements, changes in the population of settlements have influenced similar changes in the pattern of service provision. When more land is brought under irrigation, it produces more crops and it is, therefore, able to support more people and generates more wealth; the increase in crop production results in higher demand for rural services (government sector) and the increase in wealth generates more demand for retail services (private sector). The correlation coefficients are significant from another point of view; these denote a dynamic situation. The correlations are not between irrigated land, population and so on; but between changes in irrigated land, population, and so on.

The services within the taluka are located in the settlements. The distribution of settlements by service provision in 1968 was dominated by Vijayapuri and Miryalguda. In 1978, the primacy of these two towns appears to be declining. This has been made possible by a planned downward diffusion of certain services and the emergence of lower order services in the middle segment of the system (Table 13.6 and Figure 13.2).

TABLE 13.6 : **Scores of service provision and distribution of settlements 1968—78**

Centrality	Distribution of settlements					
	1968			1978		
	No.	Total score	%	No.	Total score	%
Less than 2.00	53	27.28	0.80	21	7.12	0.08
2.01 to 4.00	34	79.52	2.32	29	73.15	0.89
4.01 to 8.00	25	136.74	3.97	30	157.56	1.92
8.01 to 16.00	22	242.52	7.06	28	326.11	3.96
16.01 to 32.00	10	179.36	5.21	16	336.57	4.08
32.01 to 64.00	3	128.49	3.73	10	488.82	5.94
64.01 to 128.00	1	66.32	1.92	9	831.86	10.09
128.01 to 256.00	—	—	—	4	1,378.43	16.74
256 and above	2	2,579.66	74.99	3	4,639.32	56.30
	150	3,439.89	100.00	150	8,238.94	100.00

SERVICE CENTRES

Spatial characteristics of service centres

It was seen in the previous section that the organisation of rural and retail services in the taluka reflects an ordered adjustment to a scale of population thresholds. Some settlements have services and others do not have them; further, complexity in numbers and types of services has tended to be the basis of spatial interaction. If the service is commonly available, then the spatial interaction is on a limited scale which results in small service area. On the other hand, a less commonly available service can generate spatial interaction on a broad scale which will result in a big service area.

Four service centres were identified by an earlier study (Sen and others, 1971); these were Miryalguda, Vijayapuri, Peddavoora and Chelakurthi. The taluka, at the time of the earlier study (1968-69), was a backward, drought prone area ;

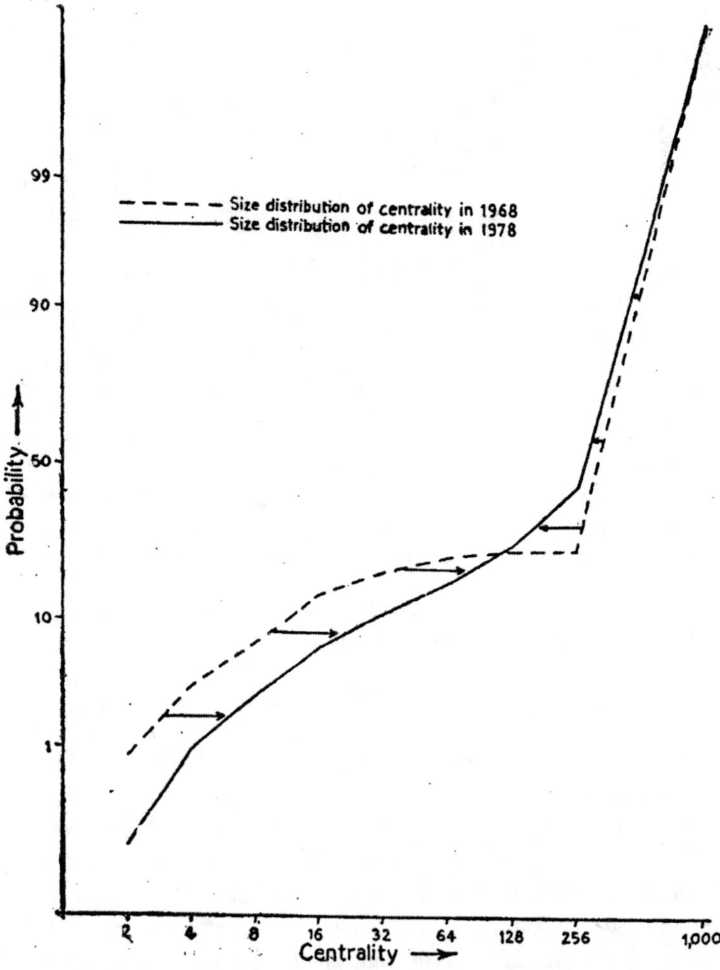

Fig. 13.2 : Miryalguda Region : Size distribution of centrality 1968-78.

its resource base was very poor ; and it had just been provided with irrigation waters. The above four centres were, incidentally, administrative headquarters of some kind. Miryalguda was a taluka and a Panchayat Samiti headquarters, Peddavoora was a Panchayat Samiti headquarters, Vijayapuri was the headquarters of Nagarjunasagar Hydro Electric Project (NSP) authority and Chelakurthi was a minor administrative headquarters (but the only one of its kind in Miryalguda Taluka) of the Left Bank Canal of NSP. The earlier study noted that this concentration of services in these four centres tended to create an imbalance in their spatial provision (Sen and others, 1971). It therefore recommended that in order to reduce the regional imbalance, a further 18 service centres needed to be 'created' ; to be operational by the beginning of the Fifth Plan period (Sen and others, 1971).

These 22 service centres were surveyed once again in 1978-79 to determine if the earlier recommendations were followed and also to note the changes in their functional and spatial characteristics. These 22 centres are : Adividevulapalle, Anjanpalle, Chelakurthi, Damercherla, Madgulapalle, Pamulapahad, Tadakamalla, Peddadevulapalle, Peddavoora, Tungapahad, Veerlapalem, Anumula, Kannekal, Marepalle, Muppavaram, Nellikal, Perur, Rajavaram, Ramadugu, Thummadam, Vijayapuri and Miryalguda.

The present analysis of service centres and their service areas and service populations attempts to note the two *changes* : (i) in the spatial incidence of rural and retail services which are located at these service centres, and (ii) service areas and service populations. Further, the changes in the interrelationships among these factors are also noted. This exercise is conducted in two stages ; first the changes are noted for each service in all service centres and second, an aggregate picture of all changes in all service centres is presented. A deeper understanding is also sought of the particular impact of services provided by the government and private sectors ; first by analysing them separately and then generalising their regional patterns. Within the government sector services, a distinction is made between those services which are simply provided by it and those whose service areas are also delineated for them ; the latter are termed 'contrived' by Johnson (NCAER, 1965 ; Johnson, 1970). In the present study, they are designated as 'regulated services' ; because their service areas are largely regulated by the government. The following are the regulated services in the taluka : veterinary hospital ; nationalised and rural banks ; co-operative banks and primary credit societies, fertiliser, pesticides and seeds distribution centres (in the government sector) and regulated markets.

The service area of each service, and all services, located in the service centres was noted and mapped. A list was made of settlements which were served by the service centres for each service and for all services. This procedure was adopted to identify the service areas of all services located in all service centres for the years 1968 and 1978. From this list, the service populations of all services located in all service centres were also obtained. The figures for the population of settlements were obtained from the 1961 and the 1971 Censuses of India, Nalgonda district.

Rural services in the taluka

There are twenty rural services in the taluka. These are all provided by the government and in some instances even their service areas are 'delineated' for them by the government.

The average area served by each one of these services in 1968 and 1978 is given in Table 13.7. The figures are interesting in two respects ; first, these show that there has been a decline in the extent of the service areas of almost all services. The three services in whose cases the service area has grown are branch post office, sub-post office and centre for allopathic treatment and clinic ; second, the higher order services have more extensive service areas and, indeed, the service areas of the lower order services 'nest' under those of the higher order. The general trend of decline in the size of the service areas is to be expected because the government has since 1968 embarked upon a plan for the location of new services, of both higher and lower orders, in the taluka. Any new service, when located in a service centre, tends to lower the average extent of the service area. Similarly, the 'nesting' of service areas is also not unexpected because various development

TABLE 13.7 : **Service area of rural services in Miryalguda taluka : 1968-78**

Services	Service area in Sq. miles	
	1968	1978
Primary school	9.90	9.90
Primary credit society	14.94	10.06
Centre for medical checkup	12.98	9.96
Bus stop/service	33.58	29.01
Branch post office	16.48	22.28
Middle school	12.83	10.12
Sub-post office	140.99	167.46
Animal husbandry centre	257.62	44.97
Secondary school	89.16	47.96
Allopathic treatment and clinic	27.08	34.15
Rural bank	297.96	91.30
Junior college	285.62	240.56
Surgery and hospital	285.62	240.56
Posts and telegraphs	285.62	240.56
Nationalised bank	571.24	240.56
Fertiliser and pesticides	728.81	728.81
Co-operative bank	728.81	728.81
Seeds distribution centre	728.81	728.81
Veterinary hospital	728.81	728.81
Regulated market	—	728.81

Source : Field Survey, 1978-79.

programmes have tended to adhere to a hierarchical system of provision of services. In such cases the areas of responsibility, and accountability, are clearly defined and 'delineated' which results in 'nesting' of the service areas.

As was noted above, it is only in the cases of branch post office, sub-post office and centre for allopathic treatment and clinic that the extent of the service areas has increased. This could, perhaps, be explained by the fact that there has been no dramatic increase in the number of settlements which have these services. (There was no increase at all in the number of service centres providing these services). Therefore, these services have gained in the service areas as more villages got used to them since 1968. There has been no change in the extent of the service areas of the highest order services such as veterinary hospital, seeds, fertiliser and pesticides distribution centres and co-operative bank because there has also been no change in the number of settlements providing these services since 1968 and also because their service area is delineated by the government to cover the entire taluka.

The service population of rural services also shows that there has been a general decline and that the service population of the lower order services tends to nest under that of the higher order services. Not unexpectedly, because of the increase in their service areas, the branch post office, the sub-post office and the centre for allopathic treatment and clinic have also gained in the size of their service populations. (Table 13.8) Some services such as veterinary hospital, seeds, fertilisers and pesticides distribution centres and co-operative

TABLE 13.8 : **Service population of rural services in Miryalguda taluka : 1968-78**

Services	Service population	
	1968	1978
Primary school	4,980	4,220
Primary credit society	6,753	4,420
Centre for medical checkup	5,832	4,822
Bus service/stop	14,257	9,087
Branch post offiee	6,298	7,259
Middle school	6,980	5,822
Sub-post office	44,240	54,920
Animal husbandry centre	57,273	14,767
Secondary school	27,856	15,539
Allopathic treatment and clinics	10,832	10,699
Rural bank	92,645	29,683
Junior college	93,283	96,826
Surgery and hospital	93,283	96,826
Posts and telegraphs	93,283	96,286
Nationalised bank	93,283	96,286
Fertiliser and pesticides	222,339	232,366
Co-operative bank	222,339	232,366
Seeds distribution centre	222,339	232,366
Veterinary hospital	222,339	232,366
Regulated market	—	232,366

Source : Field Survey, 1978-79.

bank have also registered an increase in their service populations ; but this is not because these have gained in the extent of the service areas, which they have not ; this is because of the general increase in the population of the taluka.

Retail services in the taluka

There are ten retail services in the taluka which are all provided by the private sector. In some cases, such as the provision of health and agricultural input services, the private sector appears to compete with the government. Thus, it is not uncommon to find centres for allopathic treatment and clinics and fertilisers and pesticide shops which are provided by both the government and the private sectors. There are no 'regulated' service areas for any one of the retail services.

It is interesting to note that despite a lack of a hierarchical system of responsibility and accountability, the retail services still reflect an ordered adjustment to the factor of complexity ; where it is defined as a function of population thresholds. Admittedly, there has been a greater change in the relative importance of retail services during 1968-1978 when compared to the rural services. This is clearly observed in both the figures of service area and service population (Tables 13.9 and 13.10). The services which have grown strikingly in

TABLE 13.9 : **Service area of retail services in Miryalguda taluka : 1968-78**

Services	Service area in Sq. Miles	
	1968	1978
Retail kirana store	12.67	12.67
Retail cloth store	32.23	45.59
Tea and coffee shop	10.95	10.55
Hardware store	259.89	177.52
General provision store	150.65	65.56
Fertiliser and pesticides shop	211.58	73.33
Allopathic treatment and clinic	35.35	26.35
Restaurant	155.73	120.22
Weekly market	195.06	179.84
Chemists	302.75	141.53

TABLE 13.10 : **Service population of retail services in Miryalguda taluka : 1968-78**

Services	Service population	
	1968	1978
Retail kirana store	5,414	4,220
Retail cloth store	10,395	14,060
Tea and coffee shop	6,360	4,220
Hardware store	91,992	57,638
General provision store	46,552	20,793
Fertiliser and pesticides shop	66,800	22,842
Allopathic treatment and clinic	40,625	78,956
Restaurant	49,823	43,292
Weekly market	60,537	54,863
Chemists	92,960	45,675

Source : Field Survey, 1968-78.

the settlements as outlets, have shown a decline in both area and population served. Those which have grown less markedly have 'gained' in service areas and service populations.

Spatial features of change

In 1968, with the exceptions of Miryalguda, Vijayapuri, Peddavoora and Chelakurthi, none of the service centres had more than seven rural services (out of the twenty which were considered). These seven were a mixture of lower and middle order services such as primary school, primary credit society, centre for medical checkup, bus stop/service, branch post office, middle school, secondary school and allopathic treatment and clinic. Similarly, with the exception of the above four service centres, none had more than two retail services ; which typically, were retail kirana stores and tea and coffee shops. All these services together can be defined as basic services.

These basic services do not have an identical service area and service population. For example, in both time periods, primary school, retail kirana stores and tea and coffee shops have tended to serve the population of the service centre only (where the services were located). Bus service/stop, secondary school and allopathic treatment and clinic have tended to serve an area, and a population, which is more extensive than that served by the primary school, retail kirana stores and tea and coffee shops. On the other hand, primary credit society and the centre for medical check-up have progressively abandoned their spatial links and settled down to serve the population of the service centre only.

A mere location of a new service in a service centre, or a settlement, cannot be taken to mean that, once located, it will immediately start to serve surrounding villages. This 'location' of a 'new' service also needs to be seen in its regional context. If the service is so 'new' that it is the only one of its kind in the region, then there are bound to be associated changes in the regional implications of this location. This has happened in the case of the location of regulated markets which, as a service, is 'new' in the taluka. But if the location is 'new' only at a particular service centre, or a settlement, but not in the regional context (that is, if there are such services located in other service centres or settlements), then it is quite possible that the service and the service centres, within which it is 'newly' located, will evolve new sets of regional relationships from the moment that particular service is introduced in the settlement system. The service area and service population tend to decline when such 'new' services are introduced in a region. This seems to have happened in the cases of animal husbandry centre, rural bank, allopathic treatment and clinic, general provision store, retail cloth store, hardware store, weekly market, fertiliser and pesticides shop and chemists.

As was seen earlier, changes in the regional economy and population have tended to influence the provision of rural and retail services in the taluka ; and these services, in their turn, have shaped the patterns of spatial relationships in several ways. Upon the introduction of a new service, the service centres have shown a tendency either (i) to lose service population and service area ; or (ii) to lose service population and gain service area ; or (iii) to gain service population and lose service area ; or (iv) to gain service population and service area. The following observations can be made with regard to the spatial characteristi of these changes.

(i) *The introduction of new services has resulted in the decline of service population and service area:* This is true of all types of services, except when the service is the only one of its kind (in which case, as was seen above, new space relationships emerge). This is true of both time periods and also of the dry and the irrigated tracts. For example, of the normal cases Adividevulapalle and Madgulapalle have lost service population and service area in primary credit society, retail kirana stores and centres for medical check-up because these services have also been made available in the surrounding villages. At the level of more complex services such as animal husbandry centre, rural bank, junior college, surgery and hospitals, posts and telegraphs, nationalised banks, retail cloth stores, general provision stores, hardware stores, restaurants, weekly markets, fertiliser and pesticides shops and the chemists, which in 1968 were available in Miryalguda and Vijayapuri only, one now finds a decline in service population and service area because these services are also found in a dozen more service centres.

(ii) Vijayapuri is the only example amongst the service centres which has lost its service population but has gained in service area. This is because of two interacting reasons ; first, its own population has declined from 55,300 to 16,151 and second, its services were 'opened' to the surrounding villages on the completion of the construction phase of NSP. Both have contributed quite significantly to the emergence of Vijayapuri in 1978 as a regional service centre. Until the completion of NSP, its services were primarily used by its own population ; people from the surrounding villages did not feel tied to it. After the completion of the dam, and the subsequent decline of its own population, the services in Vijayapuri could not have hoped to survive had these not been 'opened' to other villages ; which is what seems to have happened. In addition, there was a change in the perception of the surrounding villages which accepted Vijayapuri as a centre to interact with ; hence it seems to have gained in service area.

(iii) The services which were available in both time periods ; but have grown relatively rapidly over the same period, have experienced, in both the dry and irrigated tracts, a decline in their service area but an increase in their service population. Bus stop/service is an example of this type. New bus routes and a higher frequency of bus service on the established routes have resulted in a reduction in the size of the service area. The increase in the size of service population, however, is due to a general increase in the population in the taluka.

(iv) The services which were available in both time periods ; but have grown relatively slowly, have experienced, in both the dry and irrigated tracts, an increase in their service area and in their service population. The service centres with branch post offices, secondary schools, sub-post offices, centre for allopathic treatment and clinic and retail cloth stores have tended to exhibit the above spatial characteristics.

It can be stated that the changes in the spatial characteristics of rural and retail services, and of the service centres providing them, are influenced, not only by the changes in the regional economy, government investment and population but also by the number of services, their level of complexity and the time period over which a service is available in the taluka.

Service areas and service population of centres

It was seen that in the taluka the spatial organisation of the service provision is such that the service areas of the lower order services "nest" under those of the higher order services. It is felt, therefore, that the service area of the most important service located at a service centre can be considered as the area of influence of that service centre. The population of that area can also be considered as the service population of the service centre. It can be seen that the service areas and service population of other services (with are less important) located at the same centre are less than those of the most important services. The details of number of services, the extent of service area, the size of service population and the centrality scores of each service centre, for both time periods, are given in Tables 13.11 and 13.12.

In theory it is possible to say that an increase in the number of services, and in the importance of service centres, will be reflected by an increase in their total service area and service population. This is what has happened in reality.

TABLE 13.11 : **The spatial features of service centres : 1968**

Service centre	Number of services	Service area in sq. miles	Service population	Centrality score
Miryalguda	108	728.81	2,22,339	1,259.17
Peddavoora	15	166.66	87,080	36.96
Vijayapuri	160	158.28	83,345	1,320.53
Chelakurthi	23	95.12	21,645	59.41
Pamulapahad	13	34.36	10,900	27.99
Pedda Devulapalle	7	35.32	9,142	26.96
Damercherla	15	47.50	7,916	27.20
Adividevulappalle	12	53.71	7,079	18.72
Madgulappalle	9	40.30	6,617	11.52
Anjanpalle	15	28.78	5,515	23:02
Tungapahad	8	25.89	5,397	20.51
Tadakamalla	9	14.27	3,502	14.06
Ramadugu	8	26.06	3,370	7.48
Kannekal	5	17.31	3,365	12.44
Thummadam	16	14.20	3,185	32.20
Muppavaram	7	18.63	2,947	13.36
Anumula	7	16.84	2,937	16.24
Perur	10	15.83	2,643	11.92
Veerlapalem	7	17.06	2,497	8.52
Rajavaram	6	16.32	2,465	9.98
Marepalli	7	12.97	2,319	9.28
Nellikal	4	11.03	1,800	4.12

Source ; Field Survey, 1970-79.

TABLE 13.12 : **The spatial features of service centres : 1978**

Service centre	Number of services	Service area in sq. miles	Service population	Centrality score
Miryalguda	436	728.81	2,32,366	3,397.33
Vijayapuri	176	144.98	45,848	879.32
Anumula	77	195.35	44,229	224.24
Chelakurthi	29	90·66	24,385	58.88
Anjanpalle	63	47.77	16,862	122.72
Damercherla	52	74.11	15,318	94.36
Pamulapahad	16	36.94	14,447	35.20
Peddavoora	49	53.09	12 063	163.66
Adividevulapalle	55	57.78	11,741	138.95
Madgulapalle	51	35.94	10,384	59.60
Pedda Devulapalle	31	40.21	9,683	59.50
Tungapahad	43	27.24	9,185	63.72
Thummadam	72	27.15	8,017	123.04
Kannekal	18	39.97	6,991	26.16
Tadakamalla	23	18.09	6,329	74:38
Muppavaram	17	24.97	5,536	23.20
Perur	45	15.83	4,858	55.62
Nellikal	17	18.27	4,362	19.12
Ramadugu	23	19.83	4,031	25.88
Veerlapalem	26	19.82	3,860	44.64
Rajavaram	22	23.31	3,316	28.32
Marepalli	10	12.06	2,182	25.80

Source : Field Survey, 1978-79.

During the years 1968-1978, the above variables have shown that not only are these highly correlated in the taluka but that there has also been a strengthening of the values of correlation coefficients (Tables 13.13 and 13.14). The number of services, the scores of importance, the extent of service area and the size of service population have all changed, however, in a highly selective manner in the taluka. Those service centres which show low values in all variables (one standard deviation below the regional averages), in both time periods, are located in the dry tract, and/or in areas away from the Left Bank Canal, and/or have been provided with irrigation

TABLE 13.13 : **Correlation coefficients of spatial features : 1968**

	Number of services	Service area in sq. miles	Service population	Centrality score
Number of services	1.00	0.65*	0.73*	0.70*
Service area in sq. miles	0.65*	1.00	0.97	0.97
Service population	0.73*	0.97	1.00	0.93
Centrality score	0.70*	0.97	0.93	1.00

Note : *Correlation coefficients are significant at 0.1 level (N=22). All other correlation coefficients are significant at 0.01 level.

TABLE 13.14 : **Correlation coefficients of spatial features : 1978**

	Number of services	Service area in sq. miles	Service population	Centrality score
Number of services	1.00	0.95	0 96	0.98
Service area in sq. miles	0.95	1.00	0.99	0.97
Service population	0.96	0.99	1.00	0.98
Centrality score	0.98	0.97	0.98	1.00

Note : Correlation coefficients are significant at 0.01 level (N=22)

(gravity or lift) more recently. These service centres are Ramadugu, Nellikal, Muppavaram, Marepalli, Kannekal, Chelakurthi and Pamulapahad (in the dry tract) ; Veerlapalem and Rajvaram (away from the Left Bank Canal) and Tadakamalla and Devulapalle (recent introduction of canal irrigation). In addition, there is some evidence also that the service centres in close proximity to more important service centres have also grown slowly ; Tungapahad (near Miryalguda) and Perur (near Anumula) are two such examples.

In order to understand the changes in the areas of influence of service centres, the latter are ranked on the basis of the size of their service population for both time periods. Changes are noted both in upward and downward shifts in their positions. This composite picture of changes in the ranking of service centres mirrors the empirical evidence already noted in this chapter. Miryalguda excepted (and this does not mean that it has not changed at all ; in fact it has changed quite substantially, but in a 'closed' frame of reference it appears to have remained static), those service centres which have either remained 'unchanged' or have registered a 'drop' in their ranks during the years 1968-78 are all located in the dry tract and/or in the areas away from the Left Bank Canal and/or have received their irrigation waters only recently (Fig. 13.3).

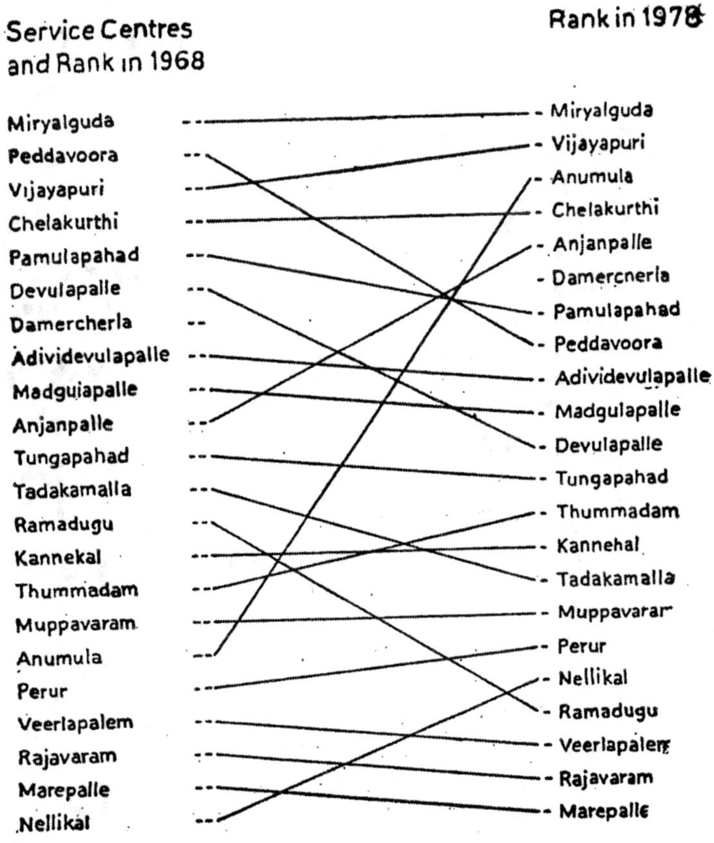

Fig. 13.3 : Miryalguda region : changes in the ranks of service centres : 1968-78

The most dramatic increase is registered by Anumula which is located at the contact zone of the dry and the irrigated tracts. It has the advantage of being located in an area halfway between Miryalguda and Vijayapuri, the two major service centres of the taluka. It is also located at the cross-roads leading to Devarkonda (west), Vijayapuri (south), Miryalguda (east) and Nalgonda (north).

Regional policies

On the basis of the evidence so far presented, it is possible, at least conceptually, to visualise a situation where due to certain structural differences the provision of services in some parts of a region is imbalanced. This calls for quick remedial measures, which take the shape of a plan to 'fill in the gaps' in the provision of services by 'creating' new service centres. It is hoped that once provided, the services will be used by the residents of the surrounding villages. It is also possible to visualise that once the government controlled services are provided in certain settlements of the region, the retail services take advantage of such locations of rural service provision by opening more retail outlets in these settlements ; and that this process of retail service provision, not controlled by the government, tends to concentrate initially in that part of the region which has such structural advantages as assured irrigation facilities, diversified cropping pattern, greater double cropping, more wealth, higher levels of rural service provision and a highly concentrated population.

The significant feature of the regional policy in Andhra Pradesh is that all the above things appear to have taken place in the taluka in 1968-1978. What is more, the horizontal spread of service centres appears to be in the process of being strengthened (with more 'policy' decisions) by vertical integration between villages and service centres. An earlier study makes some recommendations of this type and, therefore, this has not come as a surprise when the taluka was resurveyed in 1978 (Sen and others, 1971). Service areas 'defined' by government regulation are a common feature of the development administration at the district level ; there are examples of co-operative banks, regulated markets, centres for seeds and fertiliser distribution and veterinary hospitals, normally located at the headquarters of a panchayat samiti, which serve only the settlements of a panchayat samiti. What the resurvey has shown is that at the sub-panchayat samiti level also one observes these regulated/controlled service areas for certain, and new services. Thus, regulated market yards (foodgrain procurement centres), animal husbandry centres, rural banks and centres for medical treatment and clinics all show the same features (of 'defined' service areas) as do related services at a higher level. If vertical integration of rural service provision becomes more common, it will be interesting to see whether or not the retail services also 'integrate' their service areas in a similar fashion.

Further, many important changes have taken place in the settlement system of the study area. Demographically, functionally and spatially, these changes denote the emergence of a greatly altered system which even today is adjusting to the varied regional economy. Demographically, the system has become less marked by "primacy". It can be said that Miryalguda and Vijayapuri are now becoming integrated with the rural settlement system rather than being just two towns with few strong lines with their rural hinterlands. The middle segment of the settlement system has shown a tendency to fill up, thus, the shift toward a 'lognormal' system is not only as a consequence of the decline in the population of Vijayapuri but also a result of the increase in the rural population.

Functionally, the system has shown a tendency to strengthen in a hierarchical manner. Some of the rural services have followed the settlement hierarchy while being located for the first time in the study area. Higher order retail services too appear to be located in a hierarchical manner. The spatial spread of the services, both rural and retail, is outward from major centres such as Miryalguda and Vijayapuri and downward along the settlement hierarchy. That this was made possible by a policy of filling up the functional gaps in the settlement system through locating services in select centres is indicative of the success of the spatial policy of the Government of Andhra Pradesh.

Spatially, the impact of service centres has been such that it has made the services more accessible, and nearer, to the people than ever before. Service areas and service populations have generally declined over the last ten years; and this has happened to a much greater degree in the irrigated tract than in the dry tract.

It can be mentioned, on the basis of the evidence presented so far, that understanding the spatial features of a settlement system is crucial to any exercise in planning of settlement systems, about which something more needs to be done (than is being done at present) with a view to improving physical access to services that are necessary for the development of agriculture in rural areas of India.

REFERENCES

Christaller, W. (1966). *Central Places in Southern Germany*, Prentice Hall, Englewood Cliffs (New Jersey)

Johnson, E.A.J. (1970). *The Organization of Space in Developing Countries*, MIT Press (Cambridge, Mass.)

NCAER (1965). *Market Towns and Spatial Development in India*, NCAER (New Delhi).

Rao, V.L.S.P. (1961). *Some Aspects of Urbanization in South India*, Census of India, Monograph 12, a (New Delhi).

Rao, V.L.S.P., (1963). *Regional Planning*, Vikas (Bombay).

Rao, V.L.S.P., (1964). *Towns of Mysore State*, ISI, (Calcutta.)

Rao, V.L.S.P., (1964a). *Regional Aspects of Small and Medium Sized Towns of Telangana*, Research Programmes Committee, Planning Commission Report (Hyderabad.)

Rao, V.L.S.P., (1966). Urban Telangana', *Ekistics*, Vol. 21, pp. 338-41.

Rao, V.L.S.P., (1969). *Growth Promoting and Growth Retarding Factors in Development: Muzaffarnagar District—A Case Study*, Research Programmes Committee (Planning Commission) Report (New Delhi).

Rao. V.L.S.P. and Ramachandran (1974). Planning for an agricultural region, in Misra, R.P. Sundaram, K.V. and Rao, V.L.S.P. (Eds.), *Regional Development Planning in India: A New Strategy*, Vikas, (New Delhi).

Rao, V.L.S.P, and V.K. Tiwari (1975). Urban-rural India, a paper submitted to the Second Indo-British Seminar in Geography, (Cambridge).

Rao, V.L.S.P., Waheeduddin Khan and Sudhir Wanmali (1975a). *Area Planning Approach: Integrated Services Through Select Centres*,

Department of Finance and Planning, Government of Andhra Pradesh.

Rao, V.L.S.P. and V.K. Tiwari (1979). *The Structure of an Indian Metropolis, Bangalore*, Allied (New Delhi).

Sen, Lalit K. and others (1971). *Planning Rural Growth Centres for Integrated Area Development : A Study of Miryalguda Taluka*, NIRD, (Hyderabad).

Wanmali, Sudhir (1971). Ranking of settlements : A suggestion, *Behavioural Sciences and Community Development*, Vol 5, No. 2, pp. 89-96.

14

SOME ASPECTS OF SPONTANEOUS AND PLANNED AGRICULTURAL SETTLEMENT

The Present Clearing-Colonization in the ASEAN Countries of South East Asia

HARALD UHLIG

THE EXTENSION of agriculturally used and settled land—cleared from the forests or by reclamation of the swamps has been developed over a long period of time not only in the ASEAN states but over many parts of the world. There have been both, constantly and steadily growing enlargements of the cultivated areas, but frequently also entire periods of explosive and thorough extension of the former 'lebensraum' by strong movements of new 'pioneer-settlement', as the late Karl Pelzer (1945) had called it in South East Asia.

Conducting geographical field work and travelling intensely over large tracts of South East Asian countries, the author has been more and more impressed by the growing extent, pace and diversity of these settlement developments during the last two decades. He collected relevant material from own observations and interviews. The study of a fast growing literature, official and semi-official reports yielded a strong impression of the great importance of these recent changes in the socio-economic pattern, but also of the severe ecological consequences, which are caused by these recent changes in the pattern of traditional rural settlement and agriculture.

A programme initiated by the Volkswagen Research Foundation of Germany, funding research projects on present-day problems of South East Asia, finally gave the opportunity to launch a study of some areas of recent spontaneous land clearing in Thailand with a team of younger geographers from Giessen and a number of Thai colleagues and younger assistants. The results will be presented in a book in 1983-84. The following paper comprises some parts from this book, which might be of wider interest for planning and development in Asia.

The reason for directing our own field study at areas of predominantly spontaneous settlement and land-reclamation was the amazing observation that there is a wealth of printed or mimeographed publications and reports on government-directed and/or foreign-aid supported official settlement schemes but the extensive land clearing that is occurring at the same time together with undirected, true pioneer settlement is passing nearly unnoticed, be it by official reports or by scientific publications. It is clearly much more difficult to obtain reliable information or even to assess the real extent of the latter, in comparison to the official projects, which are usually well-documented and more or less exactly surveyed and delimitated. We decided to focus our main interest on the latter, not only because of the scientific interest, but also because of the strong practical importance of the undirected settlement for the entire regional and socio-economic development of the States of South East Asia.

Thus, we could hardly avoid reaching certain conclusions and making certain comparisons. These can possibly be summed up by stating the controversial question of "state-directed versus spontaneous settlement?" Hoping to focus some attention on this question and to present some research findings on the hitherto neglected spontaneous land colonisation, we would like to proceed from an originally purely scientific goal to the presentation of some arguments and theses, which might be useful for the further development policy in South East Asia.

Like the majority of the developing countries of the world, South East Asia—and the ASEAN countries in particular—are experiencing today a phase of rapid modernisation. Still, a fast growth of population and likewise a strong 'land-hunger' remain one of their major problems. Notwithstanding the impressive increases in food production during the last decades, it can only partly keep pace with that dramatic growth of population, despite the initial successes in birth control. Likewise, all attempts towards industrialization and the fast growing urbanization have not yet provided sufficient employment other than agricultural work and true alternatives of development are still widely lacking. With an agricultural population of still 60-80 per cent of the total population, the need for more land remains a dominating political problem and the opening of new agricultural areas a main goal of regional and national development.

This means conflicts in landuse, especially an ever growing pressure on the forest reserves. Their fast progressing destruction by timber extraction as well as by slash-and-burn agriculture is growing into a severe ecological hazard. Within the socio-economic sectors, financial, political and problems of property rights are added and aggravate the tasks of creating new agricultural holdings on a level well above a primitive subsistence economy together with sufficient economic prospects for the future. On the negative side, these attempts are accompanied by speculative exploitation and corruption, giving priority to fast profits, with disregard to ecological stability. South East Asia's strong ethnic and cultural pluralities provide for additional political friction, wherever foreign majorities, members of different religions etc. become involved in the land conflicts. Similarly, the principles and goals of state controlled planning and administration run frequently into conflict with the activities and needs of spontaneous land colonisation and inroads into the forests. Opening up of new areas for agriculture nearly always means the clearing of more forests. Thus, the conflicts between either the forest authorities and the spontaneous settlers (or squatters, as the authorities call them) or even between the forestry administration and the state directed land-clearing schemes, follow a predetermined course. A continuous conflict of interests between the preservation of the ecological and economical values of the forests (timber, water supply, soil conservation, climatic influences, wild life protection etc.) and agricultural expansion—or rather the need for more food and for more export-crops in favour of the national economics is hard to avoid.

The provision of new agricultural land remains therefore a major problem of planning and development. Frequently, the state-directed attempts frequently slow moving and handicapped by bureaucratic obstacles have been beaten by spontaneous pioneer clearings, but at the same time often by political and social conflicts. When these conflicts have developed into volatile political and armed conflicts, or where they have developed into either insurgent activity or into the formation of the communist underground organisations, they have gained dangerous dimensions as was the case e.g. of the Moro resistance in the Southern Philippines. They have also brought about severe ecological damage.

When the close examination of satellite photographs revealed that 57 per cent of the forest-cover of the national territory still existing in 1961 had shrunk to a mere 37 per cent in 1974 (Government of Thailand), the Thai forestry administration issued an alert that created much concern. This is, however, only one of the signals for the importance of the opening of new settlement regions balanced with a policy for the equally important protection—or even improvement and afforestation at appropriate places—of the national forest wealth!

The growing dangers of deforestation were recently brought to public attention by a study of the ESCAP (Economic and Social Commission for Asia and the Pacific of the UN), presented to government delegations of fortyone member countries at the annual conference 1982. The *Far Eastern Economic Review* (May 1982—M. Richardson) reported from this study, that Indonesia, Thailand, Malaysia and the Philippines—the four countries dealt with in this paper—are mentioned alongside with India, Nepal and

Sri Lanka as showing the highest rates of deforestation. A quotation from the report said :

> Intensified seasonal flooding with loss of lives and property, water shortages in dry seasons, accelerated erosion of agricultural land, siltation of rivers and coastal waters and disappearance of plants and animal species (most of the world's flora and fauna are in tropical regions) are but a few examples of the effects of deforestation.

It goes on stating, that Thailand, once a major exporter of timber, is now an importer, and that also exports from the Philippines have dropped dramatically in recent years. Slash-and-burn shifting cultivation by the hill people is mentioned as one of the main causes for the vanishing forests. According to our experiences and observations, I would like to give even greater weight to the causes mentioned in the second line : upland settlement by expanding, land-hungry populations, commercial logging and collection of wood for fuel !

About 70 per cent of the world's tropical wood exports are now coming from Asia ! The article goes on : "Holders of timber concessions are often partnerships between foreign firms, many of them multinational companies, and local interests that have financial links with central, state or provincial government officials" . . . "While only 10 per cent of the tropical forest is made up of trees that are commercially valuable, wasteful extraction methods cause a wider damage". According of the ESCAP report, "the rate of commercial logging practised in South East Asia alone amounts to 1 to 2.7 m ha. per year, leaving an average between a third and two-thirds of the residual trees effectively destroyed".

There are, however, many close links between excessive timber extraction and recent spontaneous—to a certain degree also planned—settlement colonisation.

In conclusion, the whole process of present clearing for new agricultural habitat in South East Asia has certainly reached a scale which may be compared with the secular importance of the medieval settlement expansion in Central and Western Europe.

The state-directed settlement schemes in South East Asia have been the subject of numerous reports of planning projects and are well-documented by publications and even more so by the so-called 'grey literature' and by the many mimeographed more or less restricted reports of various authorities and organisations. Quite a number of projects (and reports) were supported by international financial, technical and expert aid. German aid programmes as well as, for example, their literature as 'byproducts', figure prominently in Thai settlement projects and in the Indonesian "transmigrasi" programme, but British, Australian, Canadian, US American, French and other nations, aid projects have similar shares. Nearly all literature, however, is overwhelmingly fixed on the planned and state-directed sector or the settlement expansion, and is usually restricted to one single project or in the utmost to one single state. Very little has appeared, in contrast, on the equally—if not more—important phenomenon of the spontaneous (in certain cases even illegal) land clearing and settlement (if not 'squatting' in the above sense). This appeared either in the form of single case studies or in attempts to summarise the entire recent 'Landnahme' (land clearing and acquisition, be it formally legal or illegal) for any country or even over entire South East Asia. It is although a general process on a supra-national scale. It represents, furthermore, not only a specific South East Asian problem but also a worldwide one in many (tropical) countries with forest reserves and an equivalent pressure of a growing agricultural population. The problem is aggravated by dwindling size of farm holdings, rural indebtedness, unsatisfactory land reforms, need for more food and for agricultural cash—and export crops. This applies, e.g., to the Himalayan Tarai zone of India and Nepal as well as to Latin-American Indian colonization, to large parts of Africa etc.

Research into the problems of spontaneous clearing and settlement processes should most certainly be considered as being of equal importance, along with research into planned settlement schemes. Whereas the latter will primarily be concerned with an establishment of prearranged concepts—feasibility studies, selection of areas, of suitable soils and crops, improvement of cultivation-techniques, marketing, development of infrastructure and with the evaluation of its performances, the former has to reveal largely unknown problems. It is a research territory full of hard to calculate ecological, economical and social risks and even human and political conflicts and competition. Statistical data or other more or less reliable basic material is widely lacking. Stating this, already demonstrates the importance and necessity of conducting studies in this field !

Our somewhat provocative question "planned versus spontaneous settlement" is an attempt to summarise some of the experiences and to discuss relevant statements from the literature. The use of the term 'versus' is not designed to create a controversy between these two ways of land colonisation, which exist side by side, anyhow. The intention is rather to make the reader more aware of the positive and negative aspects (together with their concomitant consequences) of both types. They certainly deserve the equal attention of scientific research and of practical land policy !

After reviewing all available observations and two intensive case studies in the field, a discussion seems to be worthwhile on the pros and cons of government-sponsored versus spontaneous settlement. It starts by referring to some theses by A. Klempin (1978), who had served as a project-leader of a well-known Thai-German settlement project. He introduced his presentation of state-directed settlements on new lands in Thailand, i.e. land requiring forest clearing before cultivation can be undertaken, with a juxtaposition, "State directed versus spontaneous land settlement ?", in which he contrasts the advantages and disadvantages of both types of settlement, drawing additional support from the experiences of other authors or project leaders. In conclusion these are rendered below, together with comments of our own experiences from current investigations :

Six points in favour of 'directed land settlement' :

1. State directed settlements serve the public interest more than others.
2. They permit better protection of natural resources from ucnontrolled exploitation.

3. They allow for the selection of settlers from sections of society which correspond to national development aims in respect of distribution of work places, economic circumstances of incomes and security policies.
4. The settlers pay greater attention to instructions and rules of the authorities.
5. This type of settlement affords better opportunity to learn from trial and error.
6. It promises a higher degree of integrated development.

Against a 'directed land settlement' (Klempin lists the following disadvantages :

1. High public costs.
2. Susceptibility to political manipulation, to lack of continuity of direction and poor administration.
3. Little flexibility.
4. Reluctant repayment of credits.
5. Irregularities or deviations in the allotment of resources or on the other hand standardised farm sizes without regard to differences in the soil.

The list reveals the seasoned practitioner, and Klempin did indeed run the Saraburi Project for several years ; the author largely agrees with him but for point 5 in the section 'advantages' : "better opportunity to learn from trial and error", which seems to leave considerably greater room in the case of new spontaneous land clearances, even if there is a greater chance of giving up in subsequent degradation of a field in the event of a negative result instead of an improved second attempt.

Advantages of new spontaneous settlement are given by Klempin as :

1. Lesser expenditure of public means.
2. Smoother absorption of population growth.
3. Higher proportion of experienced farmers and enterpreneurs.

4. Better attitude in spite of the lack of capital and small administration capacity.

Disadvantages of new spontaneous settlement are :

1. Waste or destruction of natural resources (especially forests).
2. Extension of subsistence agriculture.
3. Stagnation of technology at a low level.
4. Low credit-worthiness in international financing.

Though in agreement with points 1 and 4 our own experiences require some differentiation of the consequences of points 2 and 3. In the areas under investigation subsistence economy played only a subordinate role, since the majority of even the small scale settlers plant their crops with an eye to their *Commercial saleability* (particularly in the case of maize and cassava), while products for home consumption (like dry upland rice, spices, etc) are either given limited space or not cultivated at all, being bought instead. At the same time some subsistence economies may well have been set up, especially where it is a matter of marginal additional cultivation (for example, dry upland rice in the vicinity of wet rice areas of rather too small farm sizes and high proportions of landless people) or among the meanest existences of the pioneer settlers; according to our experience, however, it is only a small percentage. Still less correct is the suggestion of subsistence economy being practiced among those quantitatively difficult to assess, but nevertheless large numbers of agricultural progressive farming methods, whose cultivated areas in the new clearances produce exclusively for the market and, what is more, for export!

The objections to point 3, stagnation of technology, are to be seen in the same context. They, too, have to be restricted to the really destitute pioneers, whereas all the clearance and cultivated areas of the entrepreneurs, and even of the somewhat more successful farmers, especially when planted to maize and cassava (not to mention the already consolidated sugarcane and pineapple plantations) are positively cultivated using modern technical means, especially by the rapidly increasing usage of tractors. The small, young central places largely consist of tractor owners, tractor workshops and tractor hire firms. They work on their own largish farms with modern technical equipment, on the one hand, and, on the other, carry out contract ploughing, thus cultivating the fields of the smaller farmers who come increasingly to recognise the financial effort as more rational than the expenditure of time and energy in traditional ploughing or even hoeing, which would in any case be unable to achieve the deep working through which appears to be favourable for the permanent utilisation of the tropical forest soils. Of course, one has to grant Klempin the fact that usually a very partial selection of techniques is applied ones which have already been well proved in practice—whereas settlements which are subsidised by the state undoubtedly offer rather more room for experiments.

In conclusion Klempin raises a number of questions, containing an evaluation of the two forms of opening up land which are to be regarded as controversial:

1. Which form achieves higher productivity and efficiency for agricultural production?
2. Which one shows greater readiness to accept innovations and is quicker to realise them?
3. Which one displays the more balanced growth?
4. Which form contributes more to the national aims of a meaningful distribution of jobs and incomes?

Again, our experiences lead us to think that the answers ought to be different. In each of the four controversial points, directed land settlement essentially depends upon equipment and infrastructure (besides integrity and efficiency of the executive organs). Striking examples may be found in the differences (and corresponding successes) among the Indonesian 'Transmigrasi'—especially in their earlier pioneer-like phases, as described in Pelzer's "Pioneer settlement in the Asiatic tropics" (1945) and in the Malaysian FELDA settlements with their very well developed infrastructure and huge investments, matched by accordingly faster and more modern emancipation of the settlers and a production orientated towards a world economy. The question, however, remains this : how profitable — measured by economic standards—are highly subsidised clearance schemes compared with the yields of rationalised large-scale plantations? From the viewpoint of reasons of state this is of secondary importance, since apart from their economic aspects settlements and the opening up of land are part of a comprehensive bundle of social, demographic and political aims. A comparison of the efficiency of 'Transmigrasi' (especially in former years, where the recent development is catching up remarkably in intensity, efficiency and investments, "FELDA" will also have to consider that the Indonesian programme, though idealistic, was only partly successful so far, having to operate with modest means and over great, inter-island distances as well as natural and human geographical differences, whereas the state directed opening up of land in Malaysia was favoured by the immediate contiguity of large land resources and relatively light population pressure and supported by massive investment and a high degree of organisational effort; their economic efficiency appears to be favourable, but their social and political effects will only become evaluable in the more distant future.

It is interesting to add some quotations from W. Collier's (1980) discussion of similar questions with regard to the development in the reclamation and settlement of swamplands for new wet-rice sawahs and combined cropping systems of rice-cum-coconut cultivation:

"If we compare the government sponsored migrants with the spontaneous migrants in the swampy land of Kalimantan, the income and the welfare of the spontaneous migrants is considerably higher than that of the transmigrants. Of course, natural selection among the spontaneous migrants means that the best have survived and prospered which has not occurred among the transmigrants. Yet, the transmigrants are at a disadvantage for the following reasons:

1. The canal system in the projects has not functioned properly. It was designed to irrigate the fields in the dry season using tidal fluctuation to force river water into the transmigrants' fields. If it had been successful,

then they would have been able to plant two rice crops per year. Almost none of these transmigrants in the swampy lands in Kalimantan and in Sumatra can plant two crops of rice per year.

2. The transmigrants' rice yields are lower than what had been expected. These yields appear to be declining because of weed, rat, and fertility problems.
3. The transmigrants are not allowed to adopt the Banjarese cropping system of rice and then coconuts.
4. The better farmers among these transmigrants are not allowed to expand their operations by purchasing the land of the less successful farmers, nor do they open the swamp forest.

Although they have a higher standard of living in these projects than if they were rural labourers in Java, they could be more successful if these constraints were removed."

Generally speaking, the consideration of various sources and experiences (see e.g. also U. Scholz, 1982) shows that true spontaneous pioneering has a strong and long established standing in Indonesia too. It is usually only somewhat concealed by the extraordinary publicity given to the government directed 'transmigration', be it in official programmes, press and propaganda as well as in the scientific literature.

Considering the experiences from Indonesia, however, the divergences in the use of the term 'spontaneous' should be remembered. The official terminology of government directed resettlement, 'transmigration' as the migration is usually directed to other islands, 'transmigrating' over the sea, from Java to Sumatra or Kalimantan, e.g.). has enacted the term 'transmigrasi spontan' for those types of resettlement, which attach additional settlers (relatives of transmigrants or friends, following from the same place of origin to the new location, with only partial government funding) to existing or just emerging, official settlement schemes. Normally, however, the term 'spontaneous' settlement, clearing, colonisation etc. has been applied to the normal sense. Its aim is to distinguish all kinds of new land reclamation (usually clearing from the forest for new agricultural use and settlement, which is based on the own initiative, working force and/or capital of the 'pioneer'-settlers or likewise entrepreneurs etc., be the acquirement of land legal or illegal at the beginning. In contrast to planned or government-directed settlement (partly supported by international aid), and even in the so-called "transmigrasi spontan" to a certain degree, there is no official support whatsoever in the true spontaneous forms of land clearing and acquirement of new settlement and agricultural areas. On the contrary, they often are opposed by government institutions (especially the forestry authorities), whilst others are tolerated or even welcomed for the sake of national economy and development. The above mentioned examples given by Collier Scholz and others from Indonesia are discussing true spontaneous pioneering too.

In Thailand, the two opposed systems are faced by the dualism of planned and spontaneous opening up of new land as it developed in the free interplay of forces, among which the former is bound to produce the socially and politically more favourable effects, as indirectly confirmed by pointed communist insurgent attacks and disruptive actions against state directed new settlements in the border districts of the Khorat Plateau, whereas the extent, economic success and pace of opening up, though in conflict with the interests of the forest economy and ecological balance—is greater in the system of spontaneous annexation of land. Both forms of new settlements contain problems which are of great significance for practical development policy and aid, as well as for scientific investigation of the developing countries.

Our own case studies of recent spontaneous land clearing areas of South East Thailand have yielded further experiences pro and contra the merits of spontaneous colonisation. The fact that two areas have been checked, situated not very far from each other, which represent entirely different types of clearing-history, agriculture and social setting, makes this still more rewarding. A true 'pioneer' settlement type (Khorat Escarpment) is opposed to the Hinterland of the fast growing port-city of Chonburi, South East Thailand, which has been characterised by massive inputs by entrepreneurs with (mainly Chinese) capital investment, larger farming units, specialised marketable production and immediate processing of products, and a more expressed social stratification. The experiences in favour or against spontaneous colonisation cover a wide range. They reach from the sad facts of a higher degree of crime, especially near the clearing 'frontier' (Thailand's so-called 'Wild East' !) over divergent examples of arising social imbalance on the one hand, but of quite good social care and relations between entrepreneurs and (migrant) labourers in other farming types (or areas). There are the merits of a highly enterprising and innovative, export-oriented modern agriculture in large as well as in medium-sized farms, partly connected to new agro-industries like sugar mills, pineapple processing etc., but also a clear disregard of forestry interests. The latter may be infringed by the tendency to expand the larger holdings continuously or by a re-emerging shifting cultivation in the ecologically valuable, remaining patches of forest on hill-tops and watersheds by smaller settlers and by labourers, to gain an extra income or to supplement their food.

The spontaneous land clearing in the Chonburi Hinterland has resulted in a remarkable and in its total effects very successful economic development. Of course, such a fast and uncontrolled growth is nearly inevitable causing certain social problems. Class differentiation, e.g., has grown, and now paid-labour-hands and smallholders face a class of large entrepreneurs, connected to further income from trade and industry. In this case, it has disrupted the former social homogeneity of the Chinese ethnic minority, which grew up from immigrated wage-earners. However, it is the dependent Thai labourers, which suffer still more from the rapid rise of a Chinese upper class. So, conflicting income-groups could well grow into such ethnic conflicts !

Increasing mechanisation connected with the loss of jobs but continued increase in productivity may alienate the groups from each other. Large-size farms have proved most productive especially for sugarcane. Owner-farmers invest their profits in labour-saving measures, whilst traders who own farms tend to speculate with their capital. Farmers' innovations aim at intensification and improvement. Traders and entrepreneurs tend to experiment with new crops, cattle-ranching etc., sometimes without proper consideration of ecological conditions. If they fail, they are quickly prepared to sell these farms.

Small farmers could be given better chances in sugarcane planting. In spite of working nearly equally productive as the large ones with hired migrant labour, they are still in a disadvantage due to the quota system. Social shortcomings on the other hand are frequently balanced by free developments of credit systems, creating pattern of interrelationships of a patron-client structure, which has by no means only negative effects. Innovative ideas, capital investment into the clearings, modernisation of the settlers, agriculture and the growth of new market centres for goods and services have been quite effective.

Summing up the pros and cons of state-directed versus spontaneous settlement, the experiences from a wide range of natural and cultural environment as well as of social, economic and political conditions all over South East Asia and more detailed from the research into the two case studies of South East Thailand seem to tend in general to a more favourable view of the spontaneous settlement, as it might be expected understandably from writings issued by government officials or of experts intensely concerned with the organisation of their planned settlement schemes.

Clearly, the risks with regard to ecological responsibility and sometimes to social justice or even in the law and order situation, including aggression and crime, and political conflict will be greater in spontaneous pioneering. Examples of a serious dependence of settlers from traders and middlemen, e.g., in the maize growing region of the Khorat Escarpment cannot be denied. Likewise, the gradual improvement of many small farmers' indebtedness by the organisation of co-operatives and marketing organisations, sometimes successful in well-conducted settlement schemes, are lacking in most spontaneously cleared regions.

A certain degree of government control, or better of government support, should ease this. The main advantages of spontaneous land settlement are the incentives given by the chances freely to invest initiatives, means and hard work with the goal to establish a promising economy and habitat as soon as possible, not being hampered by bureaucracy and various restrictions. There may occur structures of social stratification and in certain cases of social imbalance. In many cases, however, a modern division of labour and the connected socio-economic stratification, which is quite normal in other sectors of the society too, must not necessarily mean social injustice. There are many cases of a new prospering agro-economies and newly opened regions, which offer promising prospects for the 'pioneers' at various socio-economic levels, valuable diversification and extension for the national economics too.

In any case, most of the (originally, partly illegal) clearings and settlements will be more or less irreversible. In the interest of political stability, economic prosperity and development, this calls for a realistic recognition of the situation by government on all levels of administration. All possible supports by development of a sufficient infrastructure, by creating improved security and the provision of appropriate property rights should be aimed at in the national and socio-economic interest. This certainly should include efficient controls against social injustice and further ecological damage. Especially, a clear division of protected and managed forest reserves and the *de facto* agriculturally utilised areas should be an important aim. Preventing of spontaneous pioneering, however, under the prevailing pressure of population surplus 'land-hunger' and the still persisting predominance of agriculture in the national economics, would only be feasible, if government directed settlement could be raised to such speed and a standard of efficiency and attractiveness that the aim for spontaneous pioneering would be surpassed by obviously better chances !

This goal may still remain inaccessible under the given situation of national budgets. Therefore, realistic arrangements, goodwill and all possible attempts of support for both types of land reclamation appears to be the pragmatic way !

REFERENCES

Collier, W.L. (1980). "50 years of spontaneous and government sponsored migration into the swampy lands of Kalimantan : Past results and future projects", *Prisma* (Jakarta).

Klempin, A. (1978). *Contribution of State-Directed Settlement to Agricultural Development*, "A case-study of the land settlement and Thai-German agricultural development project Saraburi", South Asia Institute, Heidelberg.

Pelzer, K. (1945). *Pioneer Settlement in the Asiatic Tropics*, American Ass. of Geogr. Spec. Publ. 29, New York.

Scholz, U. (1980). "Spontane Neulanderschilebung in den Waldgebieten SO Thailands," *Wandel bäuerlicher Lebensformen in SO Asien* (ed. W. Roll, U. Scholz and H. Uhlig), Gießener Geographische Schriften.

Scholz, U (1982). "Die Ablösung und Wiederausbreitungs-Wanderfeldbaus in den südostasiatischen Tropen—Beispiele aus Sumatra and Thailand," *Süd—und Südostasien* (Festschrift zum 60. Geburtstag von H. Uhlig), Geographische Zeitschrift, Beihefte, Nr. 58, Wiesbaden.

Uhlig, H. (1979). "Geplante und spontane Neusiedlung in SO Asien," *Konflikte durch Veränderun—gen in der Raumnutzung*, Schriften des Zentrums für Regionale Entwicklungs—forschung d. Univ. Giessen, Vol. II, Saarbrucken u. Fort Lauderdale.

Uhlig, H. (ed) and R. Riethmüller, U. Scholz, Napas Sirimsamphand, A. Spaeth : "Recent pioneer settlement in SE Asia, an overview of the ASEAN countries and case studies from SE Thailand". Asien—Institute Hamburg and Giessenger Geograph. Schriften.

15

TOWARDS A MARKET CENTRE LOCATIONAL MODEL*

C. PARVATI

INTRODUCTION

IT needs no emphasis that Indian agriculture of today is getting more commercialised. The evidence for this lies in the increasing marketable and marketed surpluses even in selected crops, both food and cash crops. While agricultural production is space bound, the centres of collection and distribution of produce, the centres of distribution of inputs and the location of processing units are 'punctiform'. In the chain of production-processing-distribution, the market centre has a definite role to play as an outlet and inlet, and is one, which has all along remained disjointed, resulting in functional and spatial imbalances. Market function is not only a service function but can also provide employment and promote development, even if its scale is not comparable to that of industrial sector. A market, through its increasing market arrivals, has a multiplier effect: a quantum of 2.8 million tonnes of commodities would require 28 million gunny bags, 1,17,000 railway wagon loads or 4,67,000 lorry loads and 7 million additional individual sales transactions (Baden Hop 1969).[1]

* This paper is abstracted from the author's Ph.D thesis : *Market Centres and Spatial Development in Coimbatore-Nilgiris Region*. Mysore University, 1979 (Unpublished).

In the higher order settlements, market function is one of the functions. In the lower order settlements with an agricultural resource base, market function is the prime function. With the diversification of the regional economy, the market centre's structure is diversified. The key function of the market is to promote efficient distribution of produce, enabling the producer to get his due share from the consumer's rupee and the consumer to get his due share from the producer's rupee. Depriving both, the intermediaries take the major share, particularly if the distance between the producer and the consumer is great. The problem of market centre development and its role in spatial development focus on cluster of problems: locational, functional and organisational. The common denominator in this cluster is the non-adaptability of the obsolete and dual market structure to the changing and increasing scale of production levels and changing technology of distribution. The question is how to integrate the production and distribution systems without functional and spatial gaps and overlaps? The suggested locational model provides an answer.

THE AREA OF STUDY

Coimbatore-Nilgiris region, one of the planning regions, recognised by the Tamil Nadu State Government, is selected to work out a market locational model. The model relates only to the Coimbatore district, as Nilgiris region, with its hill-forest and plantation economy, does not fit into the irrigated and dry farming, and urban industrial landscape structure of the district. The Coimbatore district ranks first in the level of agro-industrial development among the twelve districts in the state. With a rural population density of 279 per sq. mile (1971), this agro-industrial district is one of the most urbanised in the state. It was during 1931-61 that the district recorded a high growth in population and economic development through urban industrialisation and power development. The decade 1961-71 and after, 1971-74, was characterised by growth in both urban and commercial agricultural sectors with the market cities of Tirupur and Erode developing as counters to Coimbatore. The core industry of the district is the cotton textile industry with the associated and linked industries, localised in Coimbatore agglomeration. The 804 gines, 123 ginning factories, 125 textile spinning mills with a capital of Rs. 675 million invested in the mill sector and an employment base for 1.34 million workers indicate the size of the industrial complex. The industrial sector has a major contribution to the estimated regional income of Rs. 648 million. The semi-arid climate with its dominant red and black soils, the increasing area under irrigation through various irrigation projects coupled with the most dynamic farming community who could harness the harsh environment to their advantage both in irrigation farming and dry farming contributed to both commercialisation and diversification in agriculture. As many as 26 crops are grown in the district, and all the crops enter the markets. Jowar, ragi, groundnut, sugarcane, cotton, chillies, turmeric and tobacco contribute most to regional production, and hence the importance of regional markets. Some of the markets in the district like Mettupalayam, with a transit location, have to handle the produce of the district as well as the produce of the Nilgiris district (coffee, tea, potato, fruits and forest products). The cropping pattern in the district is adapted to the climate-soil—'water budget', and the seasonal cropping rhythm is well adjusted to monsoon schedule and drought and irrigation schedule, with the result, the harvesting periods are spaced. As a consequence, the commodities arrive in markets during different months in spite of peak and relatively slack months. The farmers' responsiveness to innovations, agricultural development and package programmes, and the increasing emphasis on integrated production with distribution of agricultural produce, contributed to an increase in marketed surpluses.

The distribution of soils and the different sources of irrigation, canals, tanks and wells, is such that different types of farming have different regional patterns. Wet land paddy, sugarcane and garden crops are most localised. The cotton-groundnut-cholam cropping pattern is widely distributed over the whole plateau, though cotton is localised in the east, near Erode. This pattern of crop distribution is partly reflected in the development of markets associated with the dominance of specific crops and the commodity arrivals. Erode is a turmeric market, Tirupur is a cotton market and Pollachi is a groundnut market. Perundurai and Punjaipuliampatti are sugarcane and tobacco markets, respectively. In contrast to the cotton crop, the core crop of the region from the days of East India Company, which has reached a saturation level[2] in area and production, the groundnut which is more evenly distributed, is tending to become an important crop of the district, and is reaping maximum benefits from agricultural development programmes. In addition to Pollachi, Karur and Erode are important markets for groundnut.

The above analysis brings out four distinct zonal patterns : (i) the hill-forest plantation economy zone in the north and south. (ii) the Coimbatore-Erode urban industrial axial belt with its cotton-groundnut-cholam-turmeric concentrations. This belt is the most accessible zone to road-rail and power transmission lines. There is a well developed telecommunication link system between Coimbatore, Tirupur and Erode urban industrial and market complexes. These also have well developed extra-regional links, (iii) the northern irrigated groundnut-cholam-paddy zone, and (iv) the southern groundnut-dry farming and livestock farming zone. This four-fold zonation has to be taken into account in the development strategy of the district.

The Tamil Nadu State Marketing Board is planning to evolve an organised chain of market centres in the state. The present organisational set-up is such that there are multiple organisations, authorities, systems, and channels some of which are integrated more vertically (administration) while the others are not integrated vertically and structurally (market systems). Private, public, individual and institutional organisations further complicate the multiple, unco-ordinated, conflicting decision-making and development plans both in space and sectors.

The market channel in the district operates with producer and consumer as its two polar ends, with individual and institutional functionaries working with a number of 'interfaces' and 'interpenetration' of functions. Production, market place, functionaries, processing and consumption constitute the components of the operational framework. These components are telescoped, and hence are not always compartmentalised. The commodity and functionary channels are modified due to modern communications and information system, reducing the movement of farmers and functionaries. Mini super-markets in small market villages, mobile markets, higher order special functions in lower order places on selected days, multipurpose trips of progressive farmers to higher order centres bypassing intermediate places are the seeds of change, modifying the functional and spatial systems in the Coimbatore district.

Market centres and market types have different levels and functions, and farmers' market preferences differ with their scale of operation. Market structure and market channels differ with crops. The field, the village, agro town, market town and market city with assembly, processing and distribution, all these are important but not for all crops. Village market, an important entry point in the assembly channel, with village merchants, agents for gins and mills; agro-town, with processing units; market town and market city are important for cotton, with a pyramidal structure of infrastructural facilities. For turmeric, field sales and village and agro town levels are not important. Here only market town and market cities become important, and the crop is sent direct to these centres from the field through co-operative societies. Thus turmeric market is more institutionalised even from the field level. The crop is highly localised, the centres are few and their scale of operation is large and specialised. In contrast to cotton and turmeric, groundnut, because of its small scale in production, and the consequent small initial capital, has many petty and seasonal traders entering right from the village level.

The structure of market channel for cholam and tomato is much simpler compared with that of cotton, turmeric and groundnut. Cholam, the staple food crop of the poor, is grown mostly for domestic consumption, and is sold at the field/village and shandy levels. Tomato, grown more for city markets, shows a simple market-system-channel structure due to its perishability and high risk in storage. Herein again, consumer co-operatives handle the trade directly from the producers.

Potato, a field crop of the Nilgiris, gets the maximum market benefits through the potato co-operative markets society with its numerous primary village polling depots. Commission agents' function is localised only at Mettupalayam, the contact zone market between the hills and the plateau.

The market channel of the export-oriented plantation crops like coffee and tea is well organised with a chain of primary co-operatives lining up and a few pooling agents and the overall apex marketing board operating through open auction, channeling the produce through Coimbatore to different national markets, and international markets through the ports of Cochin, Madras and Bombay. Here the planter gets a dividend from the profits of the auction sale.

The Coimbatore district which has reached a higher level in the stage of development is characterised by dual patterns of traditional market for small farmers transporting their produce in bullock carts, and petty traders, and modern markets with truck-farmers, transacting through telex messages and multipurpose movements to higher order centres by-passing intervening opportunities. The consequence is that there is both polarisation and dispersal in space but without efficient spatial division of functions. The region has 33 market centres out of which 17 are urban centres and 16 are villages, and 30 out of 33 market centres are also shandy settlements with a very wide range of market revenue ranging from few hundred rupees as in Samathur to more than 6 lakhs of rupees as in Pollachi. Functional distances are characterised by both primacy and imbalance, particularly in market infrastructure. Coimbatore, Tirupur, Erode, Pollachi, Gobichettipalayam, Dharapuram and Udumalpet reveal these imbalances. There is no compatibility and synchronisation between population size, market-size and market and urban infrastructure. As a result, the market towns like Erode and Tirupur are over strained for their market size. The market towns like Gobi have underutilised capacity.

THE MODEL

In the following section, an attempt is made to work out a market centre location model for the district. The selection of the eight market centres in the model is based on :

(a) Regional transport accessibility from 41 points of origin (Table 15.1)

TABLE 15.1 : **Regional accessibility index based on shortest path**

Nodes	Name	Distance in kms
162	Tirupur	22,517
225	Palladam	23,020
146	Avanashi	23,611
231	Kangeyam	24,287
97	Kunnathur	25,192
191	Road Crossing (Potential Node)	25,448
155	Perundurai	25,574
252	Road Crossing (Potential Node)	26,028
288	Kundalam	26,426
143	Road Crossing (Potential Node)	26,616
69	Annur	26,989
292	Pongalur	27,581
94	Road Crossing (Potential Node)	28,132
234	Vellakoil	28,180
71	Punjaipuliampatti	28,339
113	Gobichettipalayam	29,071
214	Erode	29,158
260	Road Crossing (Potential Node)	29,305
159	Bhavani	29,724
51	Coimbatore	30,008
66	Sarcarsamakulam	30,462
291	Dharapuram	30,517

TABLE 15.1 (*Contd.*)

Nodes	Name	Distance in kms
204	Modakurichi	30,567
308	Mulanur	31,652
236	K. Paramathy	31,680
52	Tudiyalur	32,242
329	Madukkarai	32,244
131	Sathyamangalam	32,687
326	Udumalpet	32,688
220	Kodumudi	32,897
322	Pollachi	34,207
328	Madathukulam	34,720
54	Mettupalayam	35,184
320	Kinathukadavu	35,844
352	Thondamurthur	35,897
135	Andhiyur	38,951
338	Anamalai	38,968
129	Ammapet	39,742
313	Aravakurichi	40,015
240	Karur	40,599
314	Thandoni	46,046

(b) Four-fold zonal structure,
(c) Production base, and
(d) The level of urban and market infrastructure.

The regional accessibility index of Tirupur is 22,517 kilometres, indicating its central location and accessibility to all the points of origin in the district. With Tirupur the regional market city as the apex, the seven market towns (Coimbatore, Mettupalayam, Gobichettipalayam, Erode, Karur, Dharapuram and Pollachi) will function as sub-regional markets, serving their respective market hinterlands, delimited on the basis of distance and modes of transport of commodities from the points of origin.

The market regions are so delineated that there is neither overlap nor gap, functionally or spatially, and the emerging pattern is one of a commercial landscape of contiguous and nested market regions. Three distance norms are used in allocating the points of origin (Block or Panchayat Union Headquarters). The centroid of the block is treated as the point of origin from all the villages in the block to these eight markets or destination points. The distance norms[3] are :

Walking distance :	8 kms (5 miles)
Bullock cart distance :	25 kms (15 miles)
'Tempo' distance :	40 kms (25 miles)

In the allocation of blocks to the different markets, though the distance factor is taken as the key determinant, yet other factors like specialisation of a single commodity in a block, or farmer's preference for a higher order centre, are also given due weightage. For example, Sultanpet block which has a high cotton production base in the 'cotton core area', is allocated to 'cotton-market' Tirupur (38.3 kms) though it is only 33.8 kms from Coimbatore. Uthukuli is one single case which is equidistant (21.3 kms) between two markets : Tirupur and Gobi. This block is located in a triple crop zone of turmeric, groundnut and cotton, the three commodities moving in three directions : turmeric to Erode, groundnut to Gobi and cotton to Tirupur. This poses a problem in allocating it to any one of the above markets. Excluding the highly localised crops like cotton and turmeric, if the dominant crop of groundnut production is taken into account, it should be allocated to Gobi. But it is allocated to Tirupur, the regional market, because of the farmers' preference to Tirupur for multipurpose visits.

The existing levels of urban and market infrastructure for all the eight market centres are tabulated in Tables 15.2 and 15.3.

TABLE 15.2 : **Ranking of market infrastructure**
(*Existing*)

Market towns	Market size	Market types*	Function-aries	Lorry owner associa-tion	Storage	Tele-phone	Telex	Rank (cumula-tive)	Rank (final)
Tirupur	6	1	2	1	4	3	2	19	V
Coimbatore	3	1	3	1	1	1	1	11	I
Mettupalayam	1	1	4	1	3	6	0	16	III
Gobi	5	1	6	1	5	7	0	25	VI
Erode	4	1	4	1	4	2	0	16	III
Karur	1	1	1	1	4	4	2	14	II
Dharapuram	7	1	5	1	6	8	0	28	VII
Pollachi	2	1	4	1	2	5	2	17	IV

*A market centre gets one point where all the four market types are functioning simultaneously.

TABLE 15.3 : **Composite index : urban market infrastructure**

(Existing)

Administrative status	Functional score	Market service score	Accessibility index	Rank (cumulative)	Rank (final)
Tirupur	3	5	1	9	I
Coimbatore	1	1	20	22	II
Mettupalayam	6	3	34	43	VI
Gobi	5	6	16	27	III
Erode	2	3	17	22	II
Karur	5	2	41	48	VI
Dharapuram	7	7	22	36	IV
Pollachi	4	4	32	40	V

The point to be emphasised here is that the existing functional gaps have been removed by assigning equal ranks to all the sub-regional market centres, while Tirupur is ranked as the regional market city. The ranking is shown in Table 15.4.

TABLE 15.4 : **Market centres—suggested ranking**

Rank order	Market centre	Market score	Remarks
			Links
Regional market city	Tirupur	1	Intra-regional, inter-state, national and international
Market towns	Coimbatore	2	Inter-state and national
	Mettupalayam	2	Inter-state, national and international
	Gobi	2	Intra-regional and intra-state
	Erode	2	Intra-state, inter-state, national and international
	Karur	2	Intra-regional and intra-state
	Dharapuram	2	Intra-regional and intra-state
	Pollachi	2	Intra-regional, inter-state, national

The model proposes a five tier market hierarchic structure with nested functions* as shown in Table 15.5.

The market centres in the model are so located in relation to each other that by joining them, a distorted hexagonal shape emerges, with one of its triangular faces elongated and converging on Karur. Thus the triangles are not exactly equilateral. In the classical central place theory, the hexagonal form of market centres is supposed to ensure efficient division of space, enclosed by the hexagon. It may therefore be assumed that the space enclosed by these eight markets is being shared by them in such a way that the entire region is fully served. It is also envisaged that the interaction between these centres, both spatially and functionally, would be complementary and not

*These are only indicative and not exhaustive.

TABLE 15.5 : **Market facilities, services and central place functions for a market centre at a given level**

Market centre	Market facilities	Market service	Market-yard amenities
Field/ village	Field cleaning grading (preliminary) storage, bar, gas plant	—	—
Central village	Pooled storage, go-downs, credit co-operatives, market co-operatives (by growers). Input sales depot (Govt.) processing units—(e.g.) rice mill, flour mill etc., consumer markets—shandies.	Repair services, market inform-tion—relaying of	Subyard of RM
Agro town	Branch of rural bank storage	—do—	
Market town	Commercial fertiliser mining and selling depots, seed multiplication centres, processing units.	A branch of agro-industrial corporation, agricultural extension services	
Market city	Commercial grading cold storage ana refrigeration, head office of credit and market corporation, crop insurance, packing material	Market intelligence and information, repair of tractors, pumps etc. cold storage equipments etc.	
Metropolis port city	Cold storage and refrigeration expert firms, head office of banking and insurance	Market-intelligence and information, market forecasting and market planning	

Table 15.5 (Contd.)

Market centre	Communication	Transport	Welfare and social service institutions
Field/ village	Post office	Bullock-cart	Mobile weekly medical service including veterinary mobile-social service
Central village	Post and telegraphs office, telephone	Bullock-cart plus tempo, branch of lorry hire service	Rural dispensaries, family planning, maternity and child welfare, veterinary hospital, vocational school-farm training and craft skill
Agro town	—do—	—do—	A multipurpose school, hospital, veterinary hospital
Market town	Posts and telegraphs office, telephone, STD	Tempo and lorry owners, association	High school, polytechnic, hospital, breeding centre colleges
Market city	STD, teleprinter telex	Lorry owners, association	High schools, colleges, polytechnic, agricultural university, crop research centre etc. Specialised medical and veterinary services
Metropolis port city	Teleprinter, telex, STD	Head office of lorry owners' association, parcel service	University centres of technical education, specialised medical and veterinary services

competitive. The market regions of these centres are the nodal regions with market centres as the 'nodes'. The weekly market regions within each one of these may be considered as components of a single, larger region. Thus an integration in space and function is attained in the suggested model.

The model takes into account the fact that both the price spread with increasing distance between producer and consumer, and the high margin to the numerous intermediaries need to be reduced. The main emphasis of the model is on institutionalisation of transactions and reduction in the number of stages in the market systems, and in the number of intermediaries. It requires that all the producers in every village need to be registered under a producers' co-operative society. It recognises the need for division and integration of functions between organisations like regulated markets, co-operatives and *mandies*. The suggested model retains production, market types, processing, export and consumption stages in the existing marketing channel. The division of function takes the form of regulated market, dealing with regularisation of trade ; co-operative institutions dealing with credit and supply of inputs ; and mandies, dealing with goods transactions. The shandies function essentially as retail trading places, releasing the wholesale trade functions to the mandies. This is essential because of the presence of small farmers. Thus division of function is ensured. In the model, of the six intermediaries in the assembling stage, only two are retained. Likewise, in the distribution stage, instead of brokers, commission agents and wholesalers operating independently, the system suggests that the agents of processing units will take up the function of distribution. The model also takes into account the conflicts and clash of interests among the present functioning of multiplicity of organisations, and suggests a set-up of Regional Market Development Authority to integrate spatial, sectoral and infrastructural dimensions (Fig. 15.1).

DEVELOPMENT STRATEGY

The development strategy for the region relates to a 20-year period : 1971-1991.[4] Urban, rural/agricultural and industrial development sectors provide the base for the proposed strategy. The sectoral allocation of 40 per cent primary, 30 per cent secondary and 30 per cent tertiary, is indicated for 1991. The population of the region is estimated to increase from 5.25 million in 1971 to 7.7 million in 1991 adopting 1961-71 growth rate of 21.6 per cent. Industrial and commercial sectors are recognised as the main growth and development promoters.

The fourfold regionalisation[5] for agriculture development, the sixfold regionalisation[6] for regional development, the four

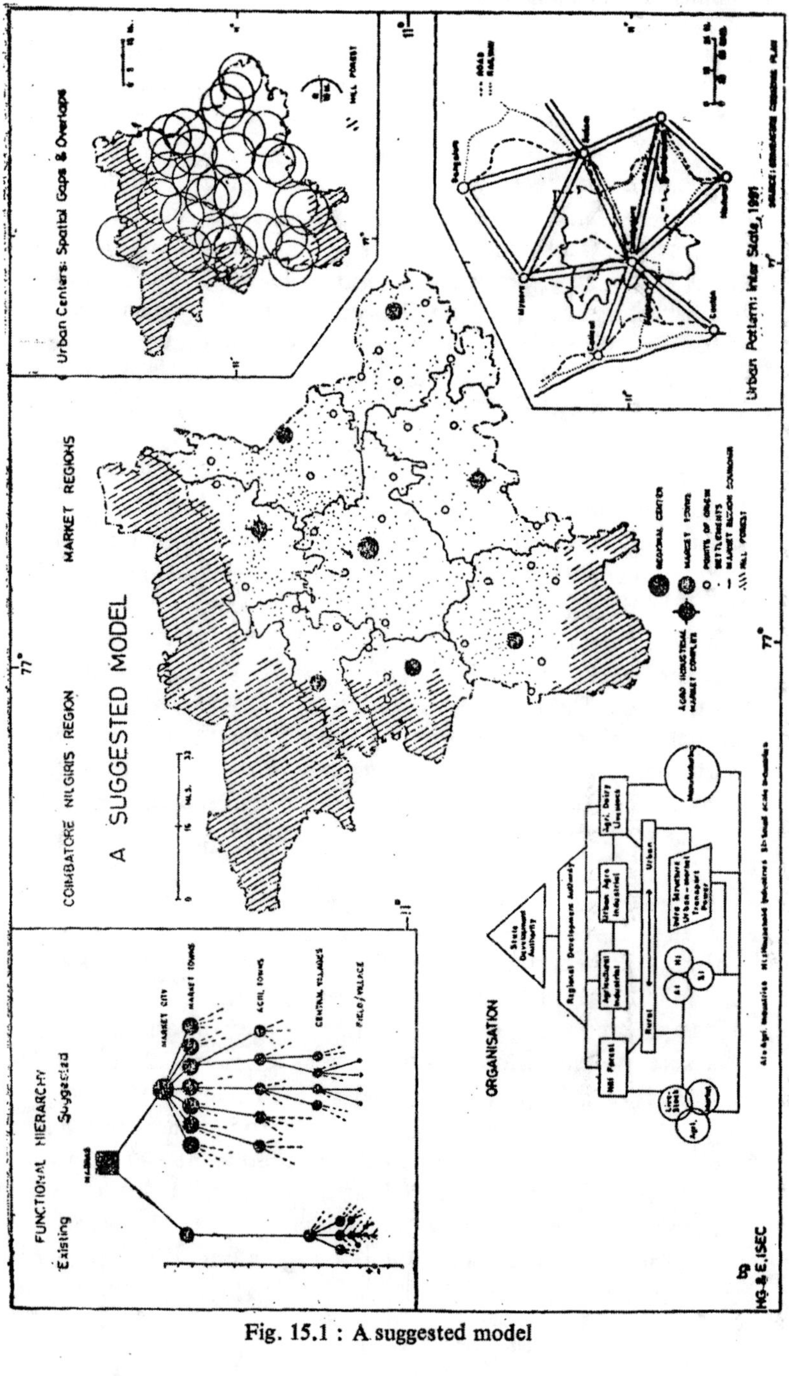

Fig. 15.1 : A suggested model

level hierarchy of settlements for the development of urban infrastructure,[7] all perpetuate the dominance of Coimbatore-Erode, and Coimbatore-Mettupalayam axial belts in the region, with Coimbatore and Erode, and also Karur and Pollachi, functioning as extra regional urban-industrial nodes. The scale of urban and market infrastructure and the agro-industrial size of the centres like Dharapuram, located in the less developed region with dairy industrial potential, and Gobi located in the resource potential region, are such that they are below the threshold level. What is needed in these centres is the location of large scale agro-industrial and market complexes based on feasibility studies. Even reconciling the regional and market development plans, the primacy of Coimbatore and the Coimbatore-Erode industrial belt persists.

"Instead of planning for the establishment of a regulated market at some place and then envisaging a 'notified area' that may be serviced by it, it is necessary to think of a region and cluster of markets suitably spaced to serve it. By simply increasing the number of regulated markets, the coverage will certainly increase but important elements in the planning of a market network will not have received adequate attention".[8] In a market development strategy, it must be recognised that market sector cannot generate employment on a scale similar to that of a large-scale industry. Market development can induce development through agro-processing industries. Market complex with appropriate market facilities generates income through market fees, sometimes of the order of Rs. 65 lakhs which could be ploughed back to provide both economic and social benefits to the farmers and traders. Market sector is more a service sector, and as such, more efficient distribution of production needs to be planned. Thus the location of the markets with reference to production base should be such that the overall transport distance is minimised, and inter-market competition is eliminated.

'The model' is oriented to the policy of 'decentralised concentration', the 'few' versus 'many'—a selective development strategy. What is needed is to strengthen a few, rather than starting new ones and keeping all of them at sub-standard level.[9]

To illustrate this, a detailed analysis for the Pollachi region is attempted. The Pollachi nodal region consists of three market centres : Pollachi, Anamalai and Kinathukadavu, where all the four types of market types exists. Anamalai and Kinathukadavu are located in the shadow of Pollachi, and are unable to develop and sustain effective market hinterlands as they are bypassed. In such a situation, both Anamalai and Kinathukadavu cannot function as independent markets though the Tamil Nadu Government has assigned independent status to these markets. At best, these centres can have pooling depots both for outputs and inputs. These centres are assigned the status of agri towns in the perspective plan for the development of market centres in this study. Anamalai with its rich resource base of forest-cum-plantation economy would become an agri town of considerable importance with the development of processing industries like saw mills, match factories, and toys and sports goods manufacturing. If horticulture is introduced along with vegetable farming on a commercial scale, the potentials are great, as there would be also extra-regional demand for these commodities just across the western border in Kerala State. Cold storage plants, fruit canning, and other related byproduct industries would create a new impulse in this tract. As a result of the Parambikulam Aliyar Irrigation Project, there is already a change in the cropping pattern. Paddy and sugarcane are increasingly cultivated with associated rice milling and jaggery making, using modern technology, thus stepping up production to supply to the extra-regional markets in Kerala State. This development would cut short a fairly long distance haulage of jaggery from Perundurai and Kavundapadi to Calicut, Trichur and Palghat.

Kinathukadavu, located midway between Pollachi and Coimbatore, on the main axial road connecting these centres, is another potential agri town because of its accessibility. Though it is a central village now, its growth is imminent, if planned development takes place in this market centre. Already, a central bus stand and a mini super-market have given a face-lift to this place. Dry farming and dairying could profitably be developed in its hinterland taking advantage the new farm technology.[10] Thus the agro towns of Anamalai and Kinathukadavu would play a complimentary role to the sub-regional market town of Pollachi. Within the hinterland of Kinathukadavu, there are 32 villages of which 5 are weekly shandies. These weekly shandies to which villagers move over a distance of 5 kms for their daily requirements could be developed as central villages.

MARKET DEVELOPMENT : COSTS AND BENEFITS

The estimated loan requirement for the development of markets in the state by the 13 market committees spread over a period of 15 years is of the order of Rs. 29.86 crores out of a total, outlay of Rs. 36.44 crores. The estimated loan requirement by the Coimbatore Market Committee is Rs. 3.4 crores[11]. As an immediate step, the Coimbatore Committee has plans to launch a "face-lift scheme" to provide basic facilities and amenities to the region's growers and traders, through a loan of Rs. 1.8 crores from Canara Bank, the region's lead bank.[12]

Expenditure of the order of crores of rupees on market development should not be a problem, considering the fact that market development is more remunerative and self-generating than community development or village industries development schemes which also consume several hundred crores of rupees (Sudhir Sen).[13] This could be debated upon, but the fact remains that investment on market development reaps direct returns through fees on commodity arrivals and licence fees for traders. The fee on commodity arrivals ranges from 25 paise to 45 paise for every 100 rupees worth of produce. It is estimated that the market committee in the state, when the project is completed, will be in a position to realise Rs. 13 crores annually on an estimated quantum of arrivals valued at Rs. 600 crores.[14] For the Coimbatore district, the estimated receipts and charges are of the order of 2.2 million rupees and 1.6 million rupees, respectively, yielding a surplus of Rs. 6 lakh in the year 1973-74. The projected estimated for the year 1981-82 are of the order of Rs. 8 million, under receipts and Rs. 7 million, under charges, with a surplus of Rs. 1 million.[15]

The Tamil Nadu market development plan proposes to spend Rs. 2.5 million rupees on 20 regulated markets and 11 minor markets in the Coimbatore district on land, construction of buildings, commercial grading and amenities and approach roads. Out of the total of Rs. 2.5 million, the development cost for grading and amenities is 21.3 lakhs. The market size as measured by the total quantum of annual commodity arrivals ranges from about 10,000 tonnes (Vallakoil) to about 86,000 tonnes (Pollachi), with development costs of 3.2 lakhs and 23.2 lakhs respectively. This is evidently the result of taking the total annual quantum of arrivals as the basis of allocating development costs (Table 15.6).

TABLE 15.6 : **Market size and development cost : Coimbatore-Nilgiris regions**

(in tonnes)

Market centres	Cost of development (Rs. in lakhs)	Total commodity (arrivals 1974)
Tirupur	13.93	56,090
Pollachi	23.24	85,880
Gobi	18.56	58,180
Erode	24.79	72,500
Coimbatore	25.02	74,570
Palladam	4.96	26,500
Perundurai	14.72	45,110
Dharapuram	16.17	49,330
Udumalpet	14.03	47,160
Annur	8.06	20,220
Punjaipuliampatti	7.10	23,520
Vellakoil	3.20	9,610
Andhiyur	12.98	41,730
Kodumudi	8.06	23,870
Kunnathur	8.07	12,680
Kinathukadavu	8.06	15,460
Thondamuthur	14.75	44,610
Anamalai	8.06	20,190
Kavundapadi	8.07	25,280
Sathiamangalam	10.07	26,810
Karur	28.10	100,166

Against the above market development costs, runs the logic of benefits, should we say hope.

One clue could be immediately found in the following example :

> For a cart load of commodity (jaggery 750 kg.), the grower receives Rs.749 in a regulated market as against Rs. 566.25 outside it. The trader actually purchases the above at a cost of Rs.754 in a regulated market as against Rs. 770.10 if it goes through *mandi* (commission agents). The reduced price to the consumer enables him to purchase more, which facilitates the farmers to increase production and the trader to handle larger volume.[16]

The other multipliers are through the development of processing industries : agro-based transport industry through the multiplication of lorries, and container industries to pack the increased production through increased incomes and inputs. This chain of sequences, if planned properly, will promote agro-services, the distribution of which again needs to be planned to eliminate the natural tendency for irregular scatter, following the present location pattern of allocation of development investment with the consequent functional and spatial gaps and/or over laps in the region.

The Tamil Nadu project plan expects that the increase in the income of the farmers will be about 15 per cent; 5 per cent from nil storage loss, 3 per cent each from premium in price and serving, consequent on the national market charges, and 2 per cent each from increased income by auction sales and market information.[17]

Against the above hope is the dupe or despair, to quote an extract from the field experience.

> From the inception of the market yards, there were neither arrivals nor sales of notified agricultural produce. The market yards at Tirupur, Coimbatore, and Palladam are practically idle after having invested about 10 lakhs of rupees on buildings, apart from establishment . . . Most

of the godowns in the Tirupur market yard are rented out to commission agents, and they are enjoying the benefits.

To the question why the growers were not utilising the market yards, the answer was that the growers were in the grip of petty dealers and moneylenders who could meet the immediate demands of the growers for discharge of their loans from banks, for manure and insecticide, seeds and even private needs. The farmers needed on the spot money by pledging their produce.

This situation became evident again and again during the field visits of the researcher, and her discussions all along the chain from producers to middlemen and consumers. The situation does not seem to be that discouraging in the case of crops like cotton and coffee, where the different links in the chain are better integrated. One way to give maximum benefits to the two polar ends of the market channel seems to be to integrate processing, distribution and production under one system of organisation, as in the case of coffee or virginia tobacco, where the marketing board or the processing units will pilot the operations all along the market channels, with appropriate sequence from inputs production, storage and distribution with a close-knit monitoring through bank credit. It must be mentioned here that the thrust for development must come more from increased income of the farmers than from the employment potential of the commercial sector. Though in terms of the numbers employed, the commercial sector in the district claims a larger number (2 lakhs)[18] compared to the industrial sector (1 lakh). Its income levels are lower than those of the industrial sector.[19]

The analysis of spatial and functional gaps in relation to production and population bases and accessibility to transport and power net, and the market centres and villages, brought out that there is no case for the development of all the 22 regulated markets and 11 minor markets. The suggested locational model indicated that the total investment of Rs. 2.5 million could be most profitably allocated to the development of the 8 markets in a phased schedule.

Thus, with the proposed model, and co-ordinating the investment on development programmes of the different departments like transport, power, agricultural marketing, urban development and other departments involved in regional development, finance should not be a bottleneck. This does not imply that the available finance for investment on the development of market complexes and market facilities are unlimited. What is needed is a phased priority investment plan for selected markets. Here, the logical step should be to take up feasibility studies indicating costs and benefits. It is necessary to stress here that a study of the structure of the market centre and its region should precede any feasibility study. The changing structure of the region, and the inter-market and market regional interrelationships need to be reviewed periodically, maybe every five years.

REFERENCES

1. Baden Hop, M.B. (1969); *Modernising the Market Structure for Foodgrains in Tamil Nadu*, General Report No. 3, USAID.
2. One view is that the cotton economy has disturbed the soil-water-table ecology to such an extent that it needs replacement and regional shift. (Discussions with the agro-economists of Tamil Nadu, Agricultural University, Coimbatore, January 1976.)
3. These distance norms were based on field enquiries in different market centres and discussions with market officials, traders and middlemen.
4. Coimbatore-Nilgiris region (1974), *A Draft Regional Plan*, pp. 63-83.
5. The Bhavani-Cauvery River Valley zone, (2) Aliyar and Thirumurthi irrigated zone, (3) Noyyal River Valley zone, (4) Southern-Central dry zone.
6. The Nilgiris-Bargur hill sub-region, (2) The Coimbatore-Tirupur sub-region, (3) The Bhavani and Cauvery River Valley sub-region, (4) The Parambikulam-Aliyar sub-region, (5) The Dharapuram-Karur sub-region, (6) The Anamalais sub-region.
7. First Order : (1) Coimbatore, Erode.

   Second Order : (1) Tirupur, (2) Ootacamund, (3) Pollachi, (4) Karur.

   Third Order : (1) Mettupalayam, (2) Dharapuram, (3) Sathyamangalam, (4) Coonoor, (5) Gobichettipalayam, (6) Udumalpet.

   Fourth Order ; (1) Kothagiri, (2) Palladam, (3) Avanashi, (4) Bhavani, (5) Kangayam. (6) Gudalur, (7) Uthukuli, (8) Annur, (9) Sivagiri (10) Kannivadi, (11) Punjaipugalur, (12) Sultanpet, (13) Karamadai,

(14) Pallapatti, (15) Periyanaickenpalayam.

8. *Market-Towns and Spatial Development in India*, pp. 41.
9. The existing 22 markets and the proposed 11 markets reveal a number of functional, spatial and organizational gaps and anomalies.
10. '*New deal for Dryland Farmers*, A comprehensive cropping programme has been evolved, taking into consideration the soil condition, rainfall and duration of new crop varieties. The programme involves fitting in of 2 short duration crops from among groundnut, millets and pulses, in the place of a single long duration crop of millets or pulses with high yielding varieties of pulses lika CO[1] Bengalgram, Anjugam, greengram, T. 9 blackgram, CO-1 redgram and high yielding millets like IS 3541, Kovilpatti tall and CSH 5 cholam (*Hindu*, dated 24.12.1976).
11. Project Report on Development of Regulated Market sin Tamil Nadu, 1973, pp. 530—537C.
12. Multi Market Complex Scheme, 1974, Coimbatore MarketC ommittee pp. 5.
13. Sudhir Sen, 1975, *Reaping the Green Revolution*, Tata McGraw Hill, New Delhi.
14. Tamil Nadu Project Report, 1975. *Ibid*, pp. 537 b and c.
15. Coimbatore Market Committee Report. Ibid.
16. Project Report, Tamil Nadu. *Ibid*. pp. 537 538.
17. Letter No. Sr. C. 1473/68 dt. 29/3/1968 from the Chairman, Coimbatore Market Committee, addressed to the Secretary to Government Agriculture Department, Govt. of Madras (quoted in Multi-Market Complex Scheme, Coimbatore 1974 Annexure I).
18. 1971 Census.
19. Absence of comparative data on income levels from different sectors does not permit any precise indication of the differences.

16

MODELLING CITY STRUCTURE FOR INTRA-METROPOLITAN ACTIVITY LOCATION

The Bangalore Experiment*

VINOD K. TEWARI

INTRODUCTION

HOW to direct the future distribution of urban population in a city and how to make the present residential areas and the newly developing residential areas in the outward moving edge of a city self-contained (in terms of neighbourhood and local service facilities) and least dependent on the city core (only for metropolitan service facilities) are two important questions posed to urban researchers and planners in the developing countries. As the city grows in population and area, a number of functions are located in residential areas and sometimes functions are shifted from the city core to the periphery, in isolation and on *ad hoc* basis; city boundary is extended step by step, residential layouts and shopping complexes are planned, and the villages within the boundary get incorporated and gradually get absorbed. What was at one time peripheral becomes an intermediate city zone. It is often not sufficiently recognized that the future city is the shifting city periphery, which is well within the "shadow zone"

*The paper is based on a study completed by the author with support from the Bangalore Development Authority (Tewari, 1978). The author is grateful to Mr. S. Jena for his assistance in data processing and computer programming for the model calibration.

of the city. Within city region different types of urban areas are identified (corporation area, urban agglomeration, conurbation area, planning area, townships etc.). There is yet no development policy for city fringe and city region. Conventional ring towns, satellites or corridors are suggested as alternative urban development patterns near the expanding cities, and the fringe developments which have already taken place ahead of the plans are regularized, and urban benefits extended.

Similarly, within the city, multistoreyed structures are constructed whether for slum dwellers or for middle and high income class households or for office-cum-shopping complexes. These sky-scraper complexes breed wherever space is available and finances are forthcoming, without any concern for the spatial distribution of residences and workplaces of people whom they are meant for; and also without any consideration of the form of the city. This pattern of *ad hoc* and haphazard development is common to all the cities in India.

The growth and development policies of urban areas are usually formulated on master plan approaches, with a considerable time lag between the preparation of the plans and their implementation. Emphasis is given to physical planning, based on unrealistic norms and standards. The interdependence between various components of the city structure is usually not taken into account. Nevertheless, the need for planning the future structure of the city based on systems approach in addition to the master plan approach is becoming more and more evident in urban planning theory and application.

It is now widely recognised that cities are embodiments of strongly inter-related elements and the complex processes by which one element of the city structure affects the others can be analysed and understood only by considering cities as urban systems. The systems approach to the study of cities (Tewari, 1983) may be defined as consisting of the following :

(a) Identification of the urban system, its elements, and the sub-systems.
(b) Developing computerised information system on

variables that characterise the elements of the urban system.

(c) Modelling urban system to reduce the complexities of the observed world to mathematical relationships and to develop predictive capabilities.

This paper is intended to illustrate the importance and rôle of urban modelling in the systems approach to the understanding, analyses and planning of the extent and locations of various activities in a metropolitan system. A city has intra-city, city-regional and inter-city dimensions, and both city-regional and inter-city dimensions contribute to the shaping of intra-city structure (Prakasa Rao, Tewari, Ramachandran 1981). However, the discussion in this paper is limited to intra-city aspects only.

URBAN MODELLING

The use of models in the analysis and planning of cities and regions started in the 1950's when large metropolitan transport studies were carried out in the United States and models of transport flows were developed. Earlier to that, some simple models of spatial interaction were in use, particularly those developed by Reilly in the 1930s. Since the mid-1960s, the time when Lowry's model (Lowry, 1964) of a metropolis was developed and high-speed electronic computers became accessible to researchers and planners, there has been an increasing interest in the use of quantitative models in urban planning.

The literature now available on urban modelling is very extensive, with a number of applications to city studies in USA and UK. In the past there were attempts to build comprehensive models with focus on a multiplicity of purposes. More recently there has been greater emphasis on simple models with limited objectives, because the comprehensive models on the one hand demand too much effort and detailed data inputs, and on the other hand become too complicated for planners. The simple ones are in fact submodels designed for modelling the subsystems of an urban system.

Models are helpful in reducing the complexities of the observed world to the coherent and rigorous language of mathematical relationships (Lowry, 1965). The theory of urban development underpinning a model represents the urban analyst's understanding or assumptions of the processes that generate a city. Considering their theoretical basis, most of the urban models can be classified as based on three approaches : the micro-economic or behaviouristic approach, the macro-approach or social physics approach, and the simulation approach.

What should be the approach for modelling the urban systems in developing countries is one question which is now being discussed widely. Are the theories of urban processes developed in the context of western cities applicable to the Indian situations ? Should we go for constructing more comprehensive models or should we focus on single problem-oriented sub-models ? Answers to these questions can be provided only by the applications of the models to various urban phenomena in different urban situations. In the context of Indian cities, very few attempts have been made in the direction of urban modelling. Although non-availability of input data and the complexities of the urban systems seem to be some of the important reasons for this, the main reason is the lack of interest in scientific approach to the analysis and planning of city structures.

The Bangalore experiment (Tewari, 1978) was one of the first attempts in the direction of modelling city structure in developing countries. Salient features of this study are presented in the following sections.

THE BANGALORE METROPOLIS MODEL

The model used in this study was based on the Lowry model of a metropolis as improved by Garin.[1] The Lowry model was developed during the period 1960-63 as part of the Pittsburgh Region to assign "urban activities to sub-areas of a bounded region in accordance with those principles of locational inter-dependence that could be reduced to quantitative form". In the model, main components in the urban structure can be described in terms of three sectors—a basic sector, a service sector, and a household sector—interacting with each other in both a spatial and a functional sense. The model consists of two gravity models, one for residential location and the other for service location, coupled together through the economic base mechanism with spatial dimension introduced in the form of site-selection and local access. Following Lowry, the three sectors are defined as :

Basic sector

This sector includes industrial, business, and administrative activities whose locations are assumed to be unconstrained by the local site selection and by problems of access to local markets ; their employment levels are primarily dependent on events outside the local economy and their clients are predominantly non-local.

Service sector

This sector comprises all those establishments—business, administrative and others, which are directly dependent on the local population and purchasing power. The site-selection of these establishments is assumed to be strongly constrained by problems of access to local residents, and employment levels are assumed to be closely related to local growth of population.[2]

Household sector

This sector consists of the resident populations whose size and location are strongly influenced by the distribution and level of employment at any given time. Also, it is assumed that residential site selection is powerfully influenced by the location of the resident's workplace.

Given the levels and locations of basic employment, the model derives its dependent population which is then allocated to different sub-areas based on an envisaged model of movement of people for work. The model then derives the requirement of the population for service activities and allocates them to service areas. Depending on the level and location of service employment, the model derives the development population on service employees and allocates them to residential areas. This, in turn, generates more demand for services which is again derived and then allocated. The model iterates in this manner until a stable solution is achieved.

The Garin-Lowry model is an extension of the original Lowry model. The developments are in the structure of the two sub-models, method of calibration and constraint procedure. The model does not treat landuse explicitly; however, this dimension can easily be incorporated into the model by translating activities into land uses, using landuse-activity ratios. Maximum and minimum capacity constraints can be imposed on the level of activity in any zone keeping in view the physical limits of the zone. Future allocation of population and employment in certain zones is usually subject to constraints because of the limited capacity of the zones to accommodate further growth. Sometimes, planning policies related to safeguarding areas with open spaces because of the environmental considerations, or to monitoring the growth of the city in particular directions will necessitate constraints on future allocation of activities.

MATHEMATICAL STRUCTURE OF THE MODEL

Notations

Z total set of zones in the study area

Z_1 set of zones in which there are no locational constraints

Z_2 set of zones in which there are only residential constraints

Z_3 set of zones in which there are only service constraints

Z_4 set of zones in which there are both residential and service constraints

m index to refer to the inner iterations of the model necessary to derive the increments of service employment and population from basic employment

n index to refer to the outer interations necessary to ensure that the locational constraints are satisfied

i, j origin and destination zones subscripts falling in the range $i, j = 1, 2, \ldots\ldots, I$

E_i total employment in zone i

Eb_i basic employment in zone i

P_j population in zone j
c_{ij} generalised cost of travel between i and j
T_{ij} number of workers employed in i and living in j
R_j residential attraction of zone j
B_j constraint weights on residential attractors R_j to satisfy residential constraints
a residential location parameter (to be calibrated)
b service location parameter (to be calibrated)
F_i service centre attraction in zone i
D_j service employees demanded by population living in j
S_{ij} number of service employees working i and demanded by population in j
Q_i constraint weights on service centre attractors F_i to satisfy service centre location constraints
p inverse activity rate (population per worker)
r population-serving ratio (service workers per capita)

At the start of the model's operations $m=1$, $n=1$, $E_i(1, n)=E_i^{*}$, $B_j(n)=1$, $Q_i(n)=1$, and all zones belong to Z_1.

Workers are distributed to their zones of residence according to

$$\text{(i)} \quad T_{ij}(m, n)=A_i(n)\, E_i(m, n)\, B_j(n)\, R_j \exp(-a c_{ij})$$

where

$$A_i(n)=1\, [\sum_j B_j(n)\, R_j \exp(-a\, c_{ij})],\ i, j \in Z$$

population living in j is found by summing equation (1) over i and applying the inverse activity rate p

$$\text{(ii)} \quad P_j(m, n)=p \sum_j T_{ij}(m, n)\ i, j \in Z$$

Service employment demanded by population living in zone j, $D_j(m, n)$, is found by applying the population serving ratio r

$$\text{(iii)} \quad D_j(m, n)=r\, P_j(m, n) \qquad i, j \in Z$$

Here, the service employment is not disaggregated as in the original Lowry model. The service employees demanded at j are distributed to various service centres according to

$$\text{(iv)} \quad S_{ij}(m, n)=K_j(n)\, D_j(m, n)\, Q_i(n)\, F_i \exp(-b\, c_{ij})$$

where

$$K_j(n)=1/[\sum_i Q_i(n)\, F_i \exp(-b\, c_{ij})],\ i, j \in Z.$$

Service employment in i is calculated by summing (4) over j.

(v) $S_i(m, n)=\sum_j S_{ij}(m, n), \qquad i, j \in Z$

and this increment of service employment becomes the input of employment to be allocated to zone of residence, according to (1), in the next interation ($m=m+1$) of the model. At this point, the preceding operations are reiterated until

(vi) $\sum_i E_i(m+1, n) < e_s$ and $\Sigma P_j(m, n) < e_p \; i, j \in Z$

where e_s and e_p are limits below which further increments of service employment and population are small enough to ignore Total population and employment (total and service) predicted by the model are

(vii) $P_j(n)=\sum_m P_j(m, n), \qquad j \in Z,$

(viii) $E_i(n)=\sum_m E_i(m, n), \qquad i \in Z,$

(ix) $S_i(n)=E_i(n)-E_i^b \qquad i \in Z.$

If there are maximum or minimum population or service employment constraints in certain zones, that is, if the set of zones z is divided into subsets Z_1, Z_2, Z_3, and Z_4, the following equations are examined. If

(x) $P_j^1 > P_j(n) > P_j^2 \qquad j \in Z_2, Z_4$

then

(xi) $P_j(n) \in Z_2^*, Z_4^*$

where P_j^1 and P_j^2 are respectively the minimum and maximum population constraints for the jth zone, and Z_2^* and Z_4^* are sets of those zones for which residential constraints have been violated,

And, if

(xii) $S_i^1 > S_i(n) > S_i^2 \qquad i \in Z_3, Z_4$

then

(xiii) $S_1(n) \in Z_3^*, Z_4^*$

where S_i^1 and S_i^2 are respectively the minimum and maximum service employment constraints for the ith zone, and Z_3^* and Z_4^* are sets of those zones for which service constraints have been violated. If equations (10) to (13) do not hold good for any zone for which constraints have been imposed, then the constraints are satisfied and the simulation terminates. However, if (10) to (13) do hold, then the coefficients B_j and Q_i are modified as

(xiv) $B_j(n+1)=B_j(n) \,.\, P_j^k/P_j(n) \quad k=1 \text{ or } 2, j \in Z_2^*, Z_4^*$
$\qquad =1 \; j \in (Z-Z^*_2-Z_4^*)$

and

(xv) $Q_i\,(n+1)=Q_i(n) \,.\, S_i^k/S_i\,(n), k=1 \text{ or } 2, i \in Z_3^*, Z_4^*$
$\qquad i \in (Z-Z_3^*-Z_4^*)$

$B_j\,(n+1)$ and $Q_j\,(n+1)$ are now substituted into (1) and (4) respectively and the whole system of proceeding equations (i) to (xv) is reiterated until the locational constraints are satisfied.

The model is calibrated at a base year, where the output of the model can be compared with the base year information. The objective of the calibration is to find the best values of the parameters *a* and *b* in the travel impedence functions in the interaction models such that the known situation is reproduced. Hyman's (1969) work provides a systematic procedure and statistic for calibrating spatial interaction models. Hyman has shown that the mean trip length is the best statistic to use in calibration. Also, it has been shown that of the various methods available for the calibration of the parameters, the method based on the Newton-Raphson numerical method provides the fastest calibration procedure. The mathematics of this procedure has been well documented in Batty and Mackie (1972), and Batty *et al.* (1973). From an initial estimate of the parameters, the Newton-Raphson method searches iteratively for improved parameter values until eventually the equation system has been solved with the estimated mean trip length equal to the actual mean trip length for journey to work and journey to services.

The calibration procedure based on Newton-Raphson numerical method was used in the Bangalore study. A description of the imput data used in the study and the results of the model calibration are presented in the following sections.

DIVISION OF THE STUDY AREA INTO ZONES

The first step in the Bangalore model building was to divide the study area (Bangalore Agglomeration, which included 63 census divisions and eight townships ; 1974 estimated population 1.94 million) into a number of geographical units to be used as sub-areas in the model. The size and shape of these units were constrained by the availability of input data required for calibrating the model. The Bangalore City Survey (Prakasa Rao and Tewari, 1979) provided the main data source for the model. These data were collected from a sample of households in the Bangalore agglomeration area during the year 1974. Although the Bangalore City Survey data were very detailed and precise those were collected from a sample designed to obtain estimates of various characteristics and inter-relationships for the city as a whole and for the three sub-divisions of core, intermediary and periphery. This had put a constraint on the number of sub-areas of the city agglomeration used for calibrating the model. It was, therefore, necessary to have lesser number of zones in order to ensure fairly good reliability of the estimates of the characteristics used in the model.

Hence, the study area was divided into fourteen zones on the basis of the following criteria :

(i) Concentration of workplaces within the zones
(ii) Homogeneity in the socio-economic characteristics of the residential population in the zones.

Data on locations of workplaces of the workers in the sample households and factorial ecology analysis of the city based on selected socio-economic, density and family characteristics were used to demarcate the boundaries of the fourteen sub-areas (Table 16.1). The eleven zones within the city corporation area were groups of the 63 census divisions; the twelfth zone was group of Jalahalli, HMT and BEL townships and Peenya and Industries on Tumkur road outside corporation limits ; the thirteenth zone included ITI townships, NGEF and Banaswadi ; and the fourteenth zone was group a of the HAL township, HAL Sanitary Board and NAL. (Fig. 16.1). The township Devarajeevanahalli and Kadagondanahalli were grouped respectively with zones 3 and 4 because of the small size of these townships and nearness to the zonal boundaries. The reason for keeping the zones as groups of census divisions and townships was to ensure the comparability of the

TABLE 16.1 : **Name of census divisions and townships included in each zone**

Zone	Census divisions and townships included in each zone
1	Yeshwantpur
2	Malleshwaram, Kodandarampuram, Swimming Pool extn. Dattatreya Temple, Seshadripuram, High Grounds Palace
3	Munireddypalayam, Deverajevenhalli
4	Frazer Town, Murphy Town, Cox Town, Kodagudanahalli
5	Joughapalayam, Ulsoor, Ashok Nagar, Shantinagar, Austin Town
6	Visveshwarapuram, Shankarapuram, Nempegowdanagar, Hanumanthanagar, Basaveswara Temple, Basavanagudi, Yediyur
7	Jayanagar, Wilson Gardens, Mavalli, Townhall
8	Binnypet, Kempapura Agrahara, Padarayanapuram, Azadnagar
9	Rajajinagar, Subramanyapuram, Maharaja Mills, Srirampuram, Bhashyamnagar, Ramachandrapuram
10	Subhashnagar, Gandhinagar, Rangaswamy Temple, Chickpet, Kashi Vishveswara Temple, K. R. Market, Kemmangudi, Super Talkies, Kalasipalyam, Dharmarajaswamy Temple, Anjanegaswamy Temple, Cubbonpet, Chamarajpet West, Chamarajpet East, Fort
11	Vidhana Soudha, St. John Hills, Swamyshetty Gardens, Commercial Street, Bharatinagar, Central Bazar, Broadway, Shivajinagar, Tasker Town, Maclver Town
12	Jalahalli, HMT, BEL Townships, Peenya and Industries around Tumkur road outside corporation limits
13	ITI, NGEF and Banaswadi
14	HAL, HAL Sanitary Board and NAL (Kodihalli)

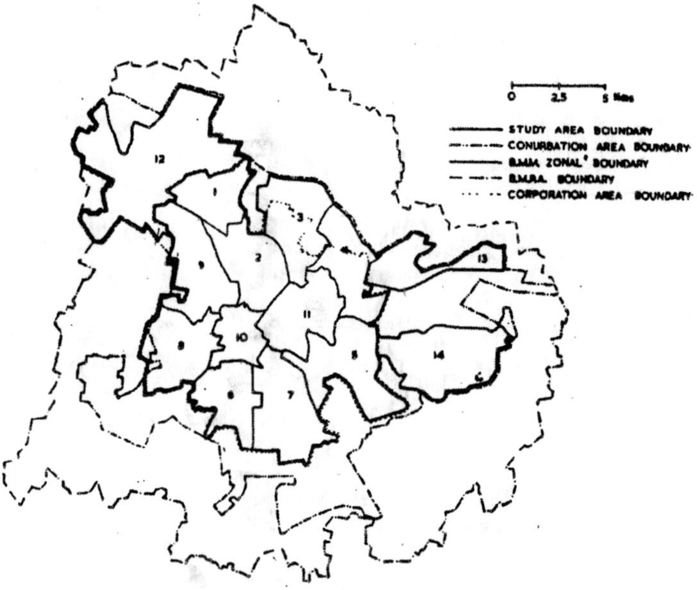

Fig. 16.1 : Bangalore metropolis model : Study area

model estimates with the secondary data available from census and other sources.

The total area covered under all the fourteen zones was 184.8 sq. kms.[5] The 1971 population in the zones (Table 16.2) varied from 10,000 (zone 13) to 273,000 (zone 10). According to the sample survey of the Bangalore Study the 1974 population of the urban agglomeration was estimated as 1.942 million. Using this figure and the 1971 figure of the urban agglomeration (1.654 million), the growth rate of the population was worked out to project the zonewise population in the year 1974.

ZONEWISE ESTIMATES OF BASIC AND SERVICE EMPLOYMENT

The most important input of the model was the break-up of total employment into basie and service categories, and their estimates by place of work for each zone. In the Bangalore City Survey, the interviews were recorded at residence with

TABLE 16.2 : **Zonewise population : 1971 (Census of India) and 1974 (estimated)**

Zone	Area in		Population in '000		Gross density per acre 1974	Residential density per acre 1974
	Acres	Sq. kms.	1971	1974		
1	2,198	8.9	35	41	19	162
2	4,446	18.0	150	176	39	178
3	1,704	6.9	55	65	38	132
4	2,841	11.5	101	118	42	131
5	4,158	16.8	124	145	35	430
6	3,236	13.1	164	183	50	261
7	3,211	13.0	134	157	49	297
8	2,890	11.7	133	159	55	350
9	3,359	13.6	212	249	74	270
10	1,457	5.6	273	320	219	2,932
11	3,063	12.4	182	214	70	680
12	7,410	30.0	30	35	5	44
13	1,828	7.4	10	12	7	51
14	3,853	15.6	49	58	15	62
Total	45,646	184.8	1,654	1,942	43	242

details of occupation, industry and place of work. The spatial distribution of workers by place of work was derived by classifying the workers by their place of work from the computerised data files of the study. In doing so, some problems arose out of the intrinsic ambiguity in the concept of place of work in some cases. For example, the place of work of the workers like autorickshaw drivers, street vendors, construction workers etc., could not be defined, as their reported place of work was "anywhere in the city". Such workers were distributed among the fourteen zones in proportion to the total number of workers working in these zones.

Information regarding occupation, industry, name of the organisation etc., was recorded in detail in the Bangalore Survey and this helped in the basic and service categorization of the employment. It was easy to determine exactly whether a worker belonged to basic or the service category. The broad groups of employment classified into the two categories are described below. Within the service classification, the employment was further classified into three sub-categories as 'neighbourhood', 'local' and 'metropolitan' services.

Basic Sector: Manufacturing ; wholesale and heavy commercial ; utilities, communications, transportation except Bangalore Transport Service and autorickshaw service ; head offices of administrative and legal services, universities and major research institutions ; head offices of financial institutions ; outdoor public services, hotels (lodging) ; mining and agriculture.

Service Sector : (a) Neighbourhood facilities : provision stores, vegetable shops ; chemists and druggists ; personal services like laundry and saloons ; lower and higher primary schools ; (b) local facilities : eating and drinking places like restaurants, hotels and bars; health and medical services—private and public ; welfare and religious services ; local branches of financial and insurance services ; department and general merchandise in retailing ; variety stores and other retail stores ; automobile and cycle repair services ; real estate services ; gasoline service stations, post offices ; secondary schools ; electricity board cash collection offices ; autorickshaw and (c) metropolitan facilities : sales agencies' offices and large photo studios ; major hospitals and colleges; theatres ; corporation, water supply and electricity board ; Bangalore Transport Service ; public libraries.

The total employment in the study area was estimated as 5,18,000, and included 275,000 basic employees and 2,43,000 service employees (Table 16.3). The basic-service employment ratio worked out to 1 : 0.886, that is, for every 1,000 basic employees there were 886 service employees.[5]

SELECTED CHARACTERISTICS OF THE WORKERS

The distribution of workers by broad income classes revealed that 50 per cent of total workers in the study area earned less than Rs. 300 per month, 44.5 per cent earned between Rs. 300 and Rs. 999, and only 5.5 per cent earned Rs. 1,000 or more (Table 16.4). The average income per worker per month was Rs. 366 and it was highest for workers working in zones 12, 13

TABLE 16.3 : **Zonewise estimated number of workers by employment type**

Zone of workplace	Basic	Service				Total
		Neigh-bourhood	Local	Metro-politan	Total	
1	3,528	340	5,114	170	5,624	8,882
2	13,940	5,270	16,830	3,074	25,174	39,114
3	3,102	170	5,130	340	2,040	5,412
4	9,886	1,374	8,810	170	10,354	20,240
5	5,624	2,734	10,100	538	13,372	18,996
6	6,134	5,780	12,410	2,890	21,080	27,214
7	28,914	3,060	16,150	5,950	25,160	54,074
8	12,240	1,530	5,440	510	7,480	19,720
9	18,034	2,380	15,680	850	18,870	36,904
10	53,904	10,200	46,948	9,520	66,668	120,527
11	41,224	4,930	27,256	8,854	41,040	82,264
12	30,590	354	1,586	170	2,110	32,700
13	19,418	170	184	0	354	19,722
14	28,196	524	3,284	0	3,801	32,004
Total	274,464	38,816	171,282	33,036	243 134	517,598

TABLE 16.4 : **Zonewise percentage distribution of workers by income classes**

Zone of workplace	Percentage of workers in monthly income class of Rs. 300	300-999	1000+	Total	Number of workers	Average income of workers
1	58.3	33.4	8.3	100.0	8,882	325
2	45.8	45.6	8.6	100.0	39,114	451
3	41.7	50.1	8.2	100.0	5,142	466
4	61.3	35.3	3.4	100.0	20,240	248
5	62.0	33.7	4.3	100.0	18,996	261
6	68.3	26.9	4.8	100.0	27,214	279
7	47.0	49.5	3.5	100.0	54,074	373
8	59.3	38.0	2.7	100.0	19,720	310
9	66.6	30.1	3.3	100.0	36,904	283
10	65.4	29.7	4.9	100.0	120,572	319
11	46.0	46.4	7.6	100.0	82,264	398
12	15.7	78.5	5.8	100.0	32,700	475
13	14.1	82.3	3.6	100.0	19,772	522
14	14.5	77.8	7.7	100.0	32,004	468
Total	50.0	44.5	5.5	100.0	517,598	336

and 14 which included the four large public sector undertakings (HMT, BEL, ITI and HAL).

While the choice of residential location in a city is constrained by a number of factors like good residential neighbourhood, nearness to shopping area, educational institutions and so on, distance to place of work plays an important role in the decision making process.[6] Therefore, the overall distribution of workers by residence-workplace separation was one of decline in the proportion of workers with increase in the distance.

Of the 517,598 workers estimated in the study area, 37 per cent travelled for work within a distance of less than a kilometre from their place of residence and 77 per cent travelled less than 5 kilometres (Table 16.5, Fig. 16.2).[7] Of those living in the peripheral township zones 12, 13, and 14, 70 to 80 per cent travelled less than a kilometre as most of the workers living in these areas worked in the same areas. Apart from township zones, the core zone 10 also had a high

TABLE 16.5 : **Zonewise distribution of workers by distance to place of work**

Zone of residence	Distance in Kms.								Total
	<1	1.0—2.9	3.9—4.9	5.0—6.9	7.0—9.9	10.0—12.9	13.0—15.9	16.0—21.0	
1	44.2	3.7	9.3	30.2	4.6	3.5	—	3.5	100.0
2	18.6	29.4	22.5	10.5	9.2	4.2	5.6	—	100.0
3	45.5	13.6	23.7	2.3	13.6	2.3	—	—	100.0
4	42.9	7.1	18.8	11.0	11,0	7.2	2.0	—	100.0
5	31.7	22.8	12.5	14.3	10.2	2.7	5.4	0.4	100.0
6	21.1	20.1	20.8	16.7	7.1	3.7	9.6	0.9	100.0
7	29.8	33.4	10.3	16.7	8.5	4.8	1.8	0.7	100.0
8	33.0	29.9	22.1	6.2	2.6	3.6	2.6	—	100.0
9	34.0	25.7	21.0	6.7	6.3	1.3	2.5	2.5	100.0
10	47.3	31.3	10.4	4.5	1.8	3.7	1.0	—	100.0
11	34.6	26.9	15.6	6.9	8.3	7.3	0.4	—	100.0
12	70.7	12.1	15.5	—	—	—	1.7	—	100.0
13	80.0	—	—	—	—	5.7	2.9	11.4	100.0
14	79.8	1.1	1.1	1.1	12.4	2.2	2.3	—	100.0
Total	37.0	24.4	15.3	9.2	6.4	3.9	3.1	0.7	100.0

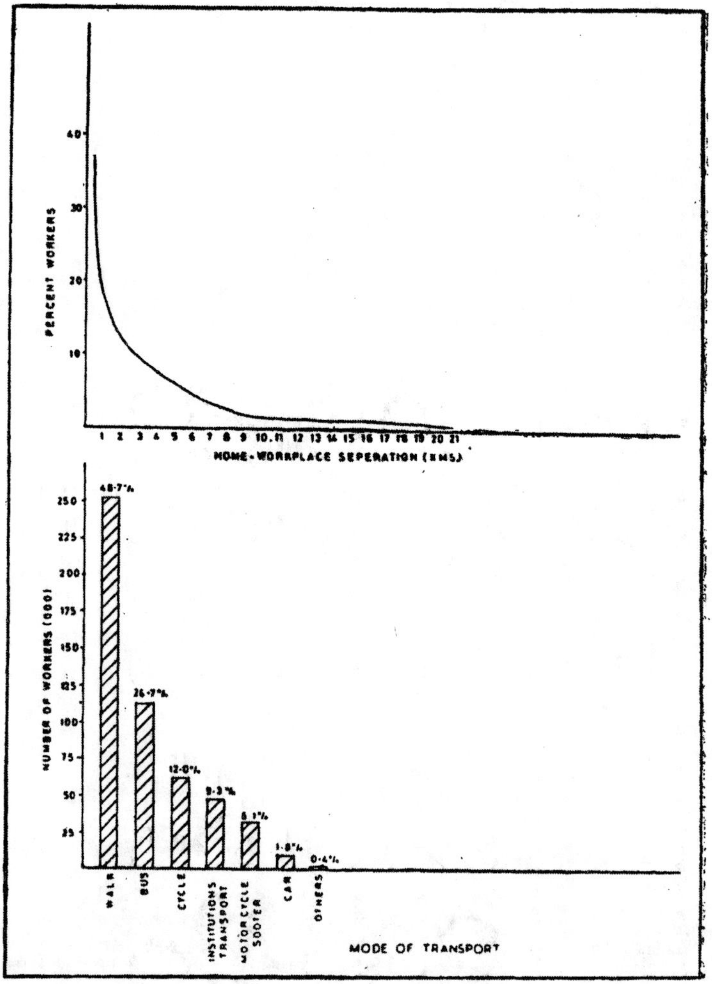

Fig. 16.2 : Movement pattern of workers by distance and mode of transport

proportion (47) of those living in the zone and travelling within one kilometre for work. Zone 2 which included high and high middle status localities like High Grounds, Palace Orchards and Malleswaram had the least proportion (19) of those travelling less than a kilometre.

The average distance travelled for work by the workers in the study area worked out to 3.3 kilometres. All the township zones had workers coming from long distances, and consequently, average distances varied from 7 to 9 kilometres. This was due to the fact that these zones included public sector undertakings which provided subsidised transport to their employees living in various parts of the city.[8]

Of the total workers (517,598), as high as 48.7 per cent travelled to their workplaces by walk (Table 16.6). Apart from

TABLE 16.6 : **Zonewise distribution of workers by mode of transport used for journey to work**

Zone of residence	Mode of transport						
	Walk	Cycle	Bus	Motor-cycle/scooter	Car	Institution's transport	Others
1	55.3	14.1	17.6	1.2	—	11.8	—
2	27.3	9.6	29.3	15.8	5.1	12.6	0.3
3	50.0	13.0	19.6	2.2	—	13.0	2.2
4	43.4	14.9	28.0	4.8	1.2	7.7	—
5	36.3	18.8	16.2	7.6	3.6	17.5	—
6	28.1	12.8	35.5	7.2	4.5	11.0	0.9
7	40.4	7.9	30.0	10.5	2.2	9.0	—
8	63.5	8.5	19.6	3.2	1.0	3 7	0.5
9	59.0	11.3	18.1	1.9	—	9.4	0.3
10	62.3	12.8	15.8	4.1	0.6	5.8	0.6
11	45.3	13.2	24.2	5.3	1.1	10.5	0.4
12	84.1	15.9	—	—	—	—	—
13	80.0	—	11.4	—	—	8.6	—
14	78.7	7.9	6.7	2.2	—	4.5	—
Total	48.7	12.0	21.7	6.1	1.8	9.3	0.4
Number of workers	252,355	61,907	112,334	31,404	9.590	47,985	1,963

walk, the most frequent mode of transport was bus (BTS) which was used by 21.7 per cent workers. This was followed, in order, by bicycle (12%), institution's transport (9.3%) and motorcycle/scooter (6.1%). Only 1.8 per cent of the workers used cars for the journey to work and 0.4 per cent used other modes of transport like, autorickshaw, train and so on.

The distribution of workers by zone of residence and by zone of workplace showed that 45 per cent of the total workers lived in the same zone where they worked (Table 16.7). Zones with important workplaces (in terms of concentration) both in terms of basic and service employment had large

TABLE 16.7 : **Zonewise number of total workers by places of residence and work and average distance travelled for work**

Zone	By place of residence		By place of work		Percentage of those working & living in the same zone to total workers by place of work
	Number of workers	Average distance (km)	Number of workers	Average distance (Km)	
1	14.790	3.6	8,882	2.6	55.5
2	54.060	4.0	39,114	2.3	49.9
3	9,348	4.1	5,142	4.4	43.8
4	27,196	3.6	20,240	1.8	67.5
5	40,970	3.8	18,996	2.1	68.0
6	56,440	4.8	27,214	1.8	60.0
7	48,450	3.3	54,074	2.5	39.6
8	33,150	2.8	19,720	2.6	38.8
9	56,270	3.1	36,904	2.4	62.2
10	92,990	2.1	120,572	2.1	47.8
11	48,790	3.2	82,264	3.0	28.7
12	11,224	1.5	32,700	7.6	33.8
13	6,624	3.8	19,772	9.1	26.1
14	17,296	2.1	32,004	7.2	43.7
Total	517,598	3.3	517,598	3.3	45.0

proportion of workers living in zones other than the zone of workplace, and also they travelled longer distances for the journey. On the other hand, zones in the intermediate areas with high proportion of residential areas and with less important places of work had small proportion of workers coming to work from outside zones.

The data on movement for activities other than work were collected, in the Bangalore Study, for twenty-two service functions. To derive the average distance travelled for service, twenty-two origin and destination matrices were prepared. These were used to derive the mean trip length for service functions, which was worked out to 2.5 kms.

CALIBRATION OF THE BANGALORE MODEL

The model for Bangalore City was calibrated using the input data derived from the data described in the previous section in a computer programme development in Basic language and operationalized on Hewlett Packard mini-computer system at the Indian Institute of Management. The following input-ratios, vectors, and matrix, were used in the model :

(i) Inverse activity rate
$$= \sum_j P_j \Big/ \sum_i E_i$$
$$= 1{,}932{,}000 / 517{,}598$$
$$= 3.73263$$

(ii) Population serving ratio
$$= \sum_i S_i \Big/ \sum_j P_j$$
$$= 243{,}134/1{,}932{,}000$$
$$= 0.12585$$

(iii) Mean work trip length
$$= \sum_i \sum_j T_{ij}\, c_{ij} \Big/ \sum_i \sum_j S_{ij}$$
$$= 3.3 \text{ kms.}$$

(iv) Mean service trip length
$$= \sum_i \sum_j S_{ij}\, c_{ij} \Big/ \sum_i \sum_j S_{ij}$$
$$= 2.5 \text{ kms.}$$

(v) Initial value of residential location parameter =2/3.3
(vi) Initial value of service location parameter =2/2.5
(vii) Zone-wise basic employment vector
(viii) Inter-zonal travel cost matrix
(ix) Zone-wise residential and service attractors

The travel cost matrix was taken as the inter-zonal distance matrix derived from the shortest path distance matrix of the transport network of the study area, except for the diagonal elements. The distances were computed between each pair of zone centroids using the Shortest Path Algorithm. The zone centroid was the geographical point of highest concentration of employment and other activities in the zone and was not necessarily the geographical centre. Since the diagonal elements in the distance matrix were zero, these were replaced, in the travel cost matrix, by the average distance of workers working and living in the same zone obtained from the survey data. These gave fairly good estimates of zone-wise intra-zonal travel distances as about 45 per cent of the total movement for work was within zones. Average distances of sample workers for journey to work between zones were not used in the travel cost matrix for the off diagonal elements because of the limited representation of the sample at that level of disaggregation.

It was pointed out in the earlier section that residential attraction of a zone is a composite function involving various factors. Similarly, the service centre attraction depends on a number of zonal characteristics. However, in this study zonewise population was used as the residential attractor and service employment as the service centre attractor.[9] The attractors were modified during the calibration depending on the constraints imposed on various zones.

With the input data described above, the model was calibrated for the year 1974. The goodness of fit of the model's simulation of the real situation of the study area in the year 1974 can be assessed from the results of the calibration presented in Tables 16.8 and 16.9 and Figs. 16.3 and 16.4.

TABLE 16.8 : **Calibration of the model, 1974**

Estimated values	Without constraints: Residential population	Without constraints: Service employment	With constraints: Residential population	With constraints: Service employment
Travel impedence parameter	0.4269	0.5762	0.4838	0.5872
Mean trip length	3.2999	2.4999	3.2914	2.5007
r-square	0.8075	0.9593	0.9288	0.9782
Absolute deviation from total in the study area	519,047	38,999	2,42,153	26,197

TABLE 16.9 : **Predicted population and service employment, 1974**

Zone	Observed values: Population	Observed values: Service employment	Calibration without constraints: Population	Calibration without constraints: Service employment	Calibration with constraints: Population	Calibration with constraints: Service employment
1	41,000	5,624	36,054	5,160	54,838	5,409
2	176,000	25,174	170,680	23,012	211,824	25,554
3	65,000	2,040	25,397	921	28,646	1,043
4	118,000	10,354	110,665	10,323	143,268	12,094
5	145,000	13,372	125,648	14,328	169,613	17,349
6	183,000	21,080	132,841	14,774	160.013	16,333
7	157,000	25,160	141,213	16,848	176,531	19,467
8	159,000	7,480	92,602	6,376	106,907	7,285
9	249,000	18,870	198,376	19,475	239,440	22,670
10	320,000	66,668	437,888	74,301	319,960*	67,244
11	214,000	41,040	276,791	42,748	213,958*	39,862
12	35,000	2,110	76,373	6,112	35,003*	2,922
13	12,000	354	17,630	375	12,000*	281
14	58,000	5,808	89,843	8,382	58,000*	5,639
Total	1,932,000	243,134	1,932,000	243,134	1,932,000	243,134

*Zones in which population contraints were applied.

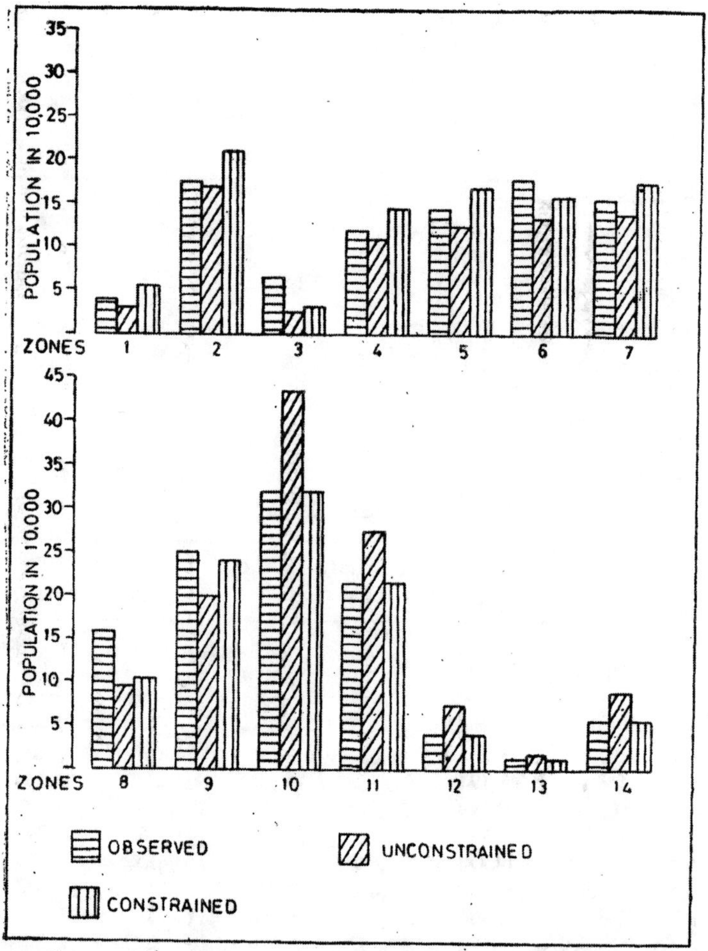

Fig. 16.3 : Observed and predicted residential population and service employment, with constraints on selected residential zones

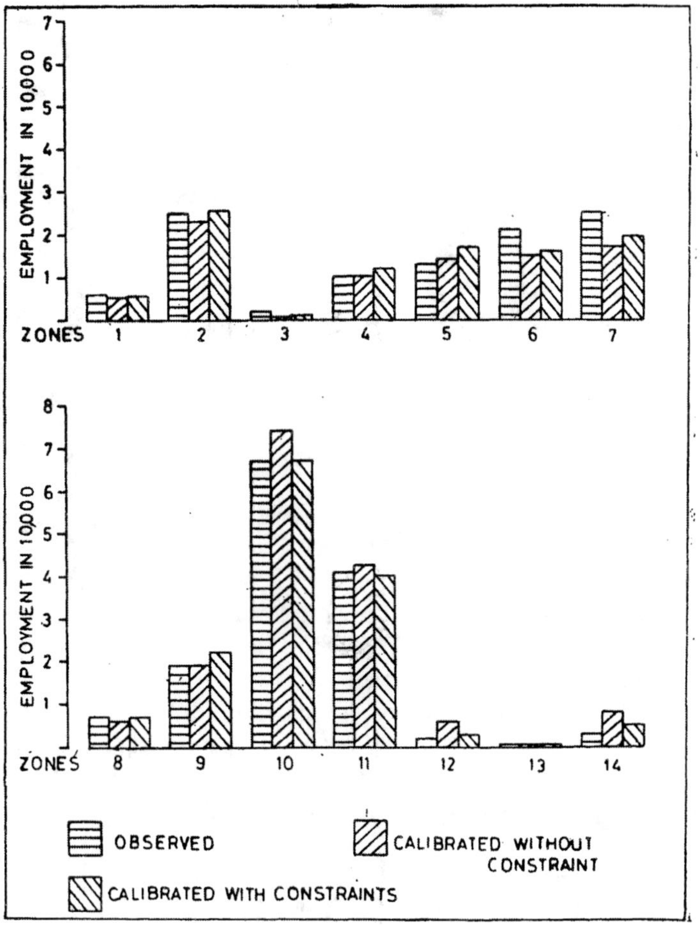

Fig. 16.4 : Model calibration : Service employment.

Calibration of the model without residential population constraints gave *r*-square value of 0.8075 for residential population.

At this point, it was necessary to find out whether the model should be calibrated with residential constraints on certain zones. A comparison of zone-wise observed and predicted values of population showed that for the core zones 10 and 11 as well as for the peripheral zones 12, 13, and 14 the predicted values were much higher than the observed values. In both the cases, it was due to higher accessibility of the zones compared to other zones. While the core zones had high accessibility due to their central locations, for the peripheral zones it was due to subsidized transport facility provided by the employers. It was, therefore, necessary to run the model with the location of residential activity constrained to the observed situation in the five core and peripheral zones.

The calibration of the model with maximum and minimum population constraints, equal to the observed population (1974), improved the goodness of fit for both residential population (*r*-square : 0.9288) and service employment (r-square : 0.9782). Thus, the constraints improved the predictive power of the model.

As an example of the application of the model, the model calibrated for the structure of the study area in 1974 was applied to simulate the future structure of Bangalore for the year 1984 based on the following assumptions :

(i) The population of Bangalore agglomeration for the year 1984 was projected to 30 lakh.[10]

(ii) The labour participation rate was assumed similar to that observed in the year 1974.

(iii) Basic-service employment ratio of 1 : 0.886 as observed in the year 1974 was assumed to change to 1 : 1 for the increased employment from the year 1974 to 1986, on the assumption that in future slightly better service facilities would be available to the residential population.

(iv) Based on assumptions (i) and (iii), the total

employment in the city in the year 1984 was estimated at about 8 lakh; an increase of 282,400 in the employment over the employment of 517,598 in the year 1974. Thus, increase in both basic and service employment was assumed to be of the order of 141,200 each.

(v) The travel cost matrix was assumed similar to that in 1974.

(vi) Further, based on the trends in the spatial spread of the city (Prakasa Rao & Tewari, 1979) it was assumed that all the increased basic employment of 141,200 would/should be located in the peripheral zones 12, 13, 14, 3, 4 and 8 : 30,000 each in zones 12, 13, and 14, and 15,000 each in zones 3 and 4, and the rest 21,200 in zone 8.

Based on the above assumptions and inputs, and using the calibrated parameter of the model (a=0.4838, b=0.5872) the city structure was simulated for the year 1984. The assumed location pattern of basic employment and the predicted allocations of residential population and service employment as determined through the model are presented in Table 16.10.

Concluding comments

The above exercise in model building, as applied to Banglore City and its future, has been one of constructing the model and demonstrating its use in the planning of intra-metropolitan activity location. It must be mentioned here that the use of any mathematical model for planning needs to be integrated with both policy goals and objectives and population targets. Thus, the role of a model is supplementary.

The Lowry-Garin model is essentially an activity allocation model. It is also a static model but the iterative procedure used in the model makes it an 'incremental dynamic' model. Dynamic versions of the model have been discussed in Crecine (1964, 1968) and Batty (1972).

TABLE 16.10 : **Predicted population and service employment, 1984**

Zone	Basic employment (assumed)	Population 1984	Service employment 1984	Increase Population	(1974-84) Service employment
1	3,258	100,448	9,722	59,448	4,098
2	13,940	338,479	40,442	162,479	15,268
3	18,102*	70,855	2,252	5,855	212
4	24,886*	303,527	23,645	185,527	13,291
5	5,624	313,816	31,755	168,816	18,383
6	6,134	227,016	22,611	44,016	1,531
7	28,914	243,503	27,099	86,503	9,939
8	33,440*	195,551	12,503	36,551	5,023
9	18,634	372,606	35,821	123,606	16,951
10	53,904	320,000**	86,886	0	20,218
11	41,224	214,000**	56,998	0	15,958
12	60,590*	70,299	5,845	35,294	3,735
13	49,418*	32,317	734	20,317	380
14	58,196*	123,612	11,227	65,612	8,119
Total	415,664	2,926,024	368,240	994,024	125,106

* Zones in which increase in basic employment was assumed.

** Zones in which existing (1974) population was used as constraint.

The Lowry-Garin model simulates the structure of a city at a time which is assumed to be an equilibrium state of the structure. Because of this the model is most useful for impact analysis (Fort, 1978). Once the model is calibrated for a city, by changing various inputs in the model, effects of future planning policy alternatives on the structure of the city could be very easily simulated. Examples of these policy alternatives are : economic policy, residential policy, service employment policy, environmental policy, and transport policy. By changing the input values of basic employment location, labour participation rates population-service employment ratio, the effects of various economic policy alternatives could be studied. Similarly, the impact of other policy alternatives could be examined by changing the locations of residential development and new services, parameters of cost of travel functions and population density and service employment constraints.

Another useful feature of the Lowry-Garin model is its flexibility with respect to level of disaggregation. The submodels could be disaggregated in terms of income and occupation groups, modes of travel-service types, and further constraints could be added, for increasing the explanatory capabilities of the model. It must be recognised that more disaggregation results in more demand on data inputs and complexities in the calibration. In the Bangalore Study, although the disaggregated data were discussed they were not used in the model calibration. It must be emphasised here that exercises in model-building require continuing work to bring out refinements in data input, in terms of spatial as well as socio-economic and behavioural data disaggregation, and model calibration.

In the context of Indian cities, there is a need for formulating spatial planning strategies for cities, city fringe and city region after a careful analysis of interrelationships between the basic entities of the systems. Urban models help planners understand these systems and provide an integrated approach to metropolitan planning.

REFERENCES

1. For a detailed description and development of the model refer to : Lowry (1964), Garin (1966), Goldner (1971), Rief (1973), and Batty (1976).
2. Lowry used the terms basic, retail and household for the three sectors, though in later studies, the terms 'service', 'population-serving' employment were used for 'retail' employment. The classification of basic employment in the Lowry model is not exactly the same as used by Hoyt (1941), Alexander (1953), Andrews (1955) and many other scholars in urban economic base analysis of cities, where the basic service split of employment is based in terms of a firm's function in relation to the regional economy (see Pfouts (1960) and Massey (1973) for detailed definitions). In the Lowry model it is the predominance of non-local and local clients and constraints in local site selection which are emphasised in the basic-service division. In this context, it is relevant to quote here Lowry (1964). "My choice of names for three groups will not please readers with a strong sense of descriptive taxonomy. It has been suggested to me that 'site-oriented' and 'residence oriented' would be more exact labels than 'basic' and 'retail', but the brief names are much more convenient in discourse".
3. In the description of the mathematical structure of the model, this section draws heavily from chapter 3 of Batty (1976).
4. Estimated from the land use map prepared by the Bangalore Development Authority. (In the Lowry model of Pittsburgh region, the study area covered 420 sq. miles and encompassed about 1.5 million inhabitants and 550.000 jobs).
5. In the Bombay Study (CIDCO 1973), two estimates of basic-service ratio were used. From Wilbur Smith Report of 1962, the ratio was estimated as 1 : 0.519, and from Traffic cell report of 1968, it was estimated as 1 : 0.813.
   In the Pittsburgh Study by Lowry, it was 1 : 0.531.
6. Answering to the question, Reason for selecting the house location, in the Bangalore City Survey, the proportion of household heads mentioning various reasons was as follows :
   Good residential neighbourhood-28 per cent ; near place of work—21 per cent ; low house rent—20 per cent ; near relatives and friends—9 per cent ; allotted house sites—7 per cent ; near shopping facilities—3 per cent ; and others—12 per cent.
7. In the Bangalore Study, the distances between the location of residence and place of work, shopping etc., were not asked by the respondent. Instead, the distance were derived from the distance matrix using the description of the locations with nearby land marks asked in the questionnaire. To derive the distance matrix, all the places in the study area were identified with the 507 nodes of the transport network in the area and distances between each pair

of nodes, i.e. the links, were measured by using a 'map measurer' from the map of the network. Using these data as input in the Shortest Path Algorithm (Rushton *et al.* 1973), the shortest actual route distance matrix was prepared. This matrix gave distance between each pair of nodes by the shortest actual route.

8. Based on an enquiry made from these undertakings in March 1975, it was found that 88 per cent of those working in HAL and BEL, 86 per cent of those working in ITI and 71 per cent of those working in HMT had their residences in the city corporation area. These undertakings provided 296 buses for the commutation of the employees at highly subsidised transport charges.
9. Experience of urban modelling of residential and service locations has shown that service employment is, in fact, a good measure of the relative attractiveness of service centres, but population is a poor measure of the relative attractiveness of residential zones.
10. The projection was based on component method (Shryock and Siegal, 1971) which involves separate projections of mortality, fertility and migration on the basis of explicit assumptions regarding each of these components of population growth. The order of migration rate was assumed as equal to the rate during 61-71. The 1981 census results showed that the rate was higher than that assumed in the projections, and the population of the agglomeration reached about 29 lakh in 1981 itself.

SUGGESTED READINGS

Alexander, J. W. (1953). An economic base study of Madison, Wisconsin, *Wisconsin Commerce Papers*, Vol. 1, No. 4, University of Wisconsin, Madison.

Andrews, R. B. (1972). Mechanics of the urban economic base, *Land Economics*, Vol. 9.

Batty, M. (1972). Recent developments in land-use modelling, *Urban Studies*, Vol. 29.

Batty, M. and S. Mackie (1972). The calibration of gravity, entropy and related models of spatial interaction, *Environment and Planning*, Vol. 4.

Batty, M. (1976). *Urban Modelling: Algorithms, Calibrations, Predictions*, Cambridge University Press, Cambridge.

CIDCO (1973). Land use planning model for Bombay, Mimeographed, City and Industrial Development Corporation (CIDCO), Bombay.

Crecine, J.P. (1964). *TOMM : Time Oriented Metropolitan Model*, CRP Technical Bulletin No. 6, Department of City Planning, Pittsburgh.

Crecine, J.P. (1968). *A Dynamic Model of Urban Structure*, P—3803, RAND Corporation, Santa Monica.

Foot, D. (1978). A computer programme for the Garin-Lowry Model, *Geographical Papers,* No. 65, Dept of Geography University of Reading.

Garin, R.A. (1966). A matrix formulation of the Lowry model for intrametropolitan activity allocation, *Journal of the American Institute of Planners*, Vol. 32.

Goldner, W. (1971). The Lowry model heritage, *Journal of the American Institute of Planners*, Vol. 37, No. 2.

Hoyt, H. (1941). Economic background of cities, *Journal of Land and Public Utility Economics*, May 1941.

Hyman. G.M. (1969). The calibration of trip distribution models, *Environment and Planning*, Vol. 1.

Lowry, I.S. (1964). *A Model of Metropolis*, RM—4035—RC RAND Corporation, Santa Monica.

Lowry, I.S. (1965). A short course in model design, *Journal of the American Institute of Planners*, Vol. 31.

Massey, D.B. (1973). The basic service categorization in planning, *Regional Studies*, Vol. 7.

Pfouts. R.W. (1960) (Ed.). *The Techniques of Urban Economic Analysis*, Chandler—Devis Publishing Company.

Prakasa Rao, V.L.S. and V.K. Tewari (1979). *The Structure of an Indian Metropolis : A Study of Bangalore*, Allied Publishers, New Delhi.

Prakasa Rao, V.L.S., V.K. Tewari and H. Ramachandran (1981). Urban system and the regional economy, *Travaux et Documents de Geographie Tropicale*, CEGET No. 45.

Reif, B. (1973). *Models in Urban and Regional Planning*, Leonard Hill Books, London.

Rushton, G. *et al* (1973) (Eds.). *Computer Programmes for Location-Allocation Problems*, University of Iowa, Iowa.

Shryock, H.S. and J.A. Siegal (1971). *The Methods and Materials of Demography*, U.S. Govt. Printing Office, Washington DC.

Tewari, V.K. (1978). Model of Bangalore metropolis, mimeographed, Institute for Social and Economic Change, Bangalore.

Tewari, V.K. (1983). Systems approach to city study, mimeographed, Indian Institute of Management, Bangalore.

17

RURALISATION OF INDIAN CITIES
A Study of the Process of Penetration of Rural Functions in Urban Areas of Slow-growing Economies

R.P. MISRA AND H.N MISRA

STUDIES on the role of urban centres in rural transformation and development are innumerable. Urbanisation and development are often seen as causally related so that the share of urban population in the total population of a country or region is usually taken as an indicator of the level of development. This is despite the empirical evidence that recent urbanisation in Third World countries has not been accompanied by industrialisation and that very many of the urban centres of these countries are nothing but overgrown villages depending for their sustenance on the tertiarisation of the regional economy.

The above premise has led many a researcher to conclude that urbanisation is a necessary condition for rural development whether it takes place through outmigration of rural people to urban centres or through the urbanisation of the rural areas. As a consequence, notwithstanding minor objections here and there, the urban expansion in the rural periphery has been accepted as a necessary evil even though it might have adverse effects on food production.

Until very recently, urbanisation in the Third World countries was not commensurate with the level of economic development. There were social factors which kept much of the rural population village-bound. With independence and the increasing impact of education, mass communication and politisation of the rural masses, the inertia that kept the people land-bound was disturbed and immigration to the urban centres, particularly the ones which had better industrial bases increased rather dramatically. Most of these centres were the large metropolitican cities. This led to what some researchers call unbalanced urbanisation.

For the first two decades after the Second World War, the primate cities of the Third World countries grew at a very rapid rate. Many of them, however, reached a point of saturation : they could not absorb any more people even in the shanty towns, favelas and slums. Quite a sizeable amount of rural population then turned its attention to intermediate size towns. One, therefore, notices an acceleration of growth of non-metropolitan towns in many Third World countries in the seventies. Next in turn will perhaps be the small towns.

The pent-up energy released after the political independence, spread of education and increased pressure of population on already dwindling agricultural resources could not sustain urbanisation for long. Most of the cities evolved a dual structure typified by formal and informal sectors of the economy. The urban society was also marked by dualism, one living in the city and the other living on the city. The latter had no stake in the city which was a means to earn a living. This gave rise to slums and favelas of the Third World countries on which enough has been written.

The brick and stone walls of the walled cities of the past were demolished to allow these cities to grow but soon a social wall was erected to keep the in-migrants away from main city. Thus each Third World city came to have the city of the 'people'.

These cities of the 'people' are both outside and inside the city of the citizens. Many of them were once outside in the periphery but have now been surrounded by new extensions. They are 'islands' within the city proper, while many others are in the periphery growing in parallel with the planned suburbs. In some cases they are relocated outside the city in order that they do not 'pollute' the city and its citizens.

Since major investments in manufacturing and social capital in Third World countries have gone to the large cities, some marginal benefits have gone to the migrants living in popular settlements too. The national and local governments also took care to ward off the 'ill-effects' of poor migrants on the city life : they were tolerated and enabled to earn their living but never allowed to take over the city. The metropolitan cities continued to encroach on the surrounding rural areas incorporating the rural settlements into their fold. In due course, the rural functions disappeared and the peasants became the proletariat and joined the bandwagons of the new migrants.

The city has spread its tentacles far longer and wider and drastically influenced and changed the rural style of life in the periphery. The migrants to city maintain a close link with the place of their origin and periodically send not only money but also the more easily emulatable urban traits of life. They are the agents of social change—a change which further accelerates rural-urban migration.

The job and work opportunities for the new migrants to city are dwindling fast but the economic and social pressure in the rural areas for the young to leave are increasing faster. This is particularly so in the stagnant or very slow-growing national and regional economies. One, therefore, notices a new trend in the rural-urban migration pattern. That is, the people move to the city along with the rural functions in which they were engaged earlier. This gives them two distinct advantages : they earn from the city without paying for its services and they have no job problem. This phenomenon is very clearly visible in the second-level cities of countries like India where the urban governments are weak to resist the pressure from the periphery and where opportunities for bringing rural functions in urban areas are better, both from the points of view of supply of land and demand for goods and services produced.

In some of the well established cities the pressure of the rural functions encroaching on the urban areas is so high that even the posh localities are now infested with goats, cows, buffaloes and pigs. The wide roads, the parks, and open areas are no more the places where people go ; they are converted into cattle sheds. In many localities the people and the animals live under the same roof, as they still do in the rural areas.

As these functions enter the block, the rental value of the houses goes down partly because the sanitary conditions deteriorate and partly because no one wants to have a tiff with the ruralites. Gradually, a larger area is in the hands of these rural migrants and this goes on expanding till they reach a point beyond which they cannot go because the land and property values are beyond their reach.

Thus one finds a unique phenomenon of the city urbanising the rural areas while at the same time the village ruralising the urban areas. One is expanding outwards while the other inwards. While earlier the villages incorporated within the municipal limits could be recognised only in terms of housing, streets and levels of social services, Now they have gone in the background and this phenomenon of rural functions invading the city is manifest almost everywhere leading to a new pattern of rural-urban relations and interaction.

The purpose of this paper is to report the results of an exploratory empirical study carried out in Allahabad—an old intermediate town of India. The city is located in the central part of the Uttar Pradesh state and has been a major educational and cultural centre of India until 1950. It was the provincial capital of the then United Provinces in the British days and was called the Oxford of the East from the educational angle.

As most other cities of India, Allahabad too had two 'cities' within one : the old city where the Indians lived and the new city where the British and the officials who carried the burden of the British Raj lived. After the independence of the country in 1947, the distinctions between the two got blurred but structural differences remain. The old is old. Its narrow streets, typically Indian architecture, the intermingling of functions are counterpoised with the planned network of roads and western-style bungalows with wide open spaces in the new. The latter also has the cantonment.

The city is located in a poverty-stricken region. Its rural hinterland is pressurising it too hard to change its character.

It can no more boast of being the Oxford of the East, nor can it take pride in its planned Civil Lines. All along the roads that fork out from the city, Allahabad is expanding in the rural periphery. But there is clear and visible reverse trend : the rural culture is penetrating the city from without leading to the ruralisation of the urban landscape.

It is a city of migrants

The population of the city (Table 17.1) has quadrupled during the last 80 years (1901-1981).[1] The population of the city is likely to be 0.80 million by 1991. The Allahabad Master Plan, however, projects to 1.26 million.

TABLE 17.1 : **Growth of population : Allahabad city (1901-81)**

Census year	Population	Variation	Per cent variation
1901	1,72,032	—	—
1911	1,71,697	−335	−0.2
1921	1,57,220	−14,477	−8.4
1931	1,83,914	+26,694	+17.0
1941	2,60,630	+76,716	+41.7
1951	3,32,295	+71,695	+27.5
1961	4,30,730	+98,435	+29.6
1971	5,13,036	+82,306	+99.1
1981	6.42.420	+1,29,384	+25.2

Source : Census Handbook of Allahabad, 1961, 1971, 1981.

Whereas earlier much of the population growth was accounted for by natural increase, if most recent and projected figures are any indicators, it is the in-migration which is spurring the population growth now (Table 17.2).

TABLE 17.2 : **Population growth components : Allahabad city (1941-91)**

Decade	Birth rate %	Death rate %	Natural increase %	Migration rate %	Total growth rate (%)
1941-51	29.00	19.5	9.5	18.00	+27.50
1951-61	26.20	9.3	16 90	12.72	+29.62
1961-71	23.50	6.5	17.00	15.00	+32.00
1971-81	20.70	5.5	15.20	30.00	+45.20
1981-91*	18.00	4.5	13.50	40.00	+53.50

Source : Allahabad Master Plan, 1967-1991, Vol. 1.

* Projected.

Poor economic base : The role of informal sector

As stated earlier, the economic base of the city is very weak. The production-oriented functions are few that too came into existence during the last two to three decades. The tertiary sector gives maximum employment i.e. 84.1 per cent (Table 17.3). And within the tertiary sector, other services which largely consist of the informal services accounted for 43.2 per cent of the workers in 1911 (Table 17.4).

TABLE 17.3 : **Occupational structure of Allahabad city (1921-81)**

(Per cent of workers)

Particulars	1921	1951	1961	1971	1981
Primary	9.4	4.0	3.7	5.7	6.7
Secondary	14.6	23.3	24.6	22.2	9.2 (includes household industry manufacturing processing, servicing, repairs)
Tertiary	76.0	72.7	71.7	72.1	84.1

Source : Census of Allahabad District, 1921-81.

Note : Decline in the share of secondary sector in 1981 due to change in definition.

TABLE 17.4 : **Distribution of workers in tertiary sector of Allahabad city (1921-71)**

Particulars	1921	1951	1961	1971
Commerce and trade	14.1	18.3	16.6	19.6
Transport and storage	7.0	10.9	12.8	9.3
Other services	53.9	43.5	42.4	43.2

Source : Census of Allahabad District, 1921-71.

Detailed figures for 1971 are not yet available but those wh o have observed the changes in the occupational structure of Allahabad during the last few years are of the view that the informal sector now accounts for about 50 per cent of the total employment in the city. It is interesting to note that the share of the primary sector in the total employment in the city has gone up from 3.7 per cent in 1961 to 7.6 per cent in 1981.

The predominance of the informal sector in the urban economy is playing an important role in the ruralisation process mentioned earlier. Every year there is a huge influx of unskilled labourers from the rural hinterland. Most of them come for a few months and they maintain their roots in the village.[2] These persons do all sorts of odd jobs such as rickshaw pulling, shoe repairing, vegetable vending, cart pulling and so on. A study carried earlier revealed that there were 40,300 rickshaw pullers during 1979-80 plying 80,850 cycle rickshaws in the city. Another study in one locality of the city registered a variety of functions in which the newcomers to the city were involved (Table 17.5).

TABLE 17.5 : **A survey of informal sector in Katra market of Allahabad city (1983)**

Particulars	Numbers	Per cent
Permanent vegetable shops	13	4.3
Roadside temporary vegetable shops	31	10.2
Thelas with seasonal vegetables	52	17.2
Thelas carrying seasonal fruits	23	7.6
Thelas selling tea, biscuits, chat, etc.	32	10.6
Betel shops	42	13.9
Footpath hosiery shops	23	7.6
Water wheel	12	4.0
Cycle repair shops	7	2.3
Plastic goods and footwear shops	19	6.3
Shoe repairers	5	1.7
Meat and fish shops	7	2.3
Miscellaneous (thela and shops)	37	12.0
Total	303	100.0

Source : Fieldwork, April, 1983.

LIVESTOCK IN URBAN AREAS : SPATIAL DISTRIBUTION

Increasing population and almost shrinking economic base of the city hence led to the influx of rural functions in the urban economy. The people are on the move : but they cannot get the opportunities in the city. They, therefore, move with whatever opportunities they have in the village. To illustrate this new trend, a study of livestock industry in the city was carried out in May-August, 1983.

Cows, buffaloes, pigs and donkeys are a common sight in Allahabad. They are almost everywhere : whether it is the collectorate, the High Court, the hospital, the Civil Lines or the bungalows in the posh localities, these animals roam everywhere. Cow dung or excreta of cattle is found even on the most widely travelled roads and thoroughfares, what to say of lanes and bylanes of the city. Even the planned sections of Civil Lines[3] are not spared by the roaming cattle and buffaloes. They thus add a new dimension to the growing informal sector of the city. The animals are concentrated in certain sectors of the city and all these sectors are slummish in appearance and remind one of the environment of the village transplanted in the city.

There are 45 slum areas in the city. And each slum area is infested with animal population. Detailed studies were carried out in five wards of the city. Some of the wards have minor slums while the others do not. But the animals are found almost everywhere. The wards selected for the survey were Rajapur, Shiv Kuti, Tagore Town, Mumfordganj and Civil Lines—all located on the north of the Northern Railway in the administrative-cum-residential zone of the city. In this zone, animals are not allowed according to municipal rules. Table 17.6 gives the livestock population of each of these wards.

On the south of this railway line is the old congested section of the city where the presence of cattle and buffaloes may not be uncommon. Main characteristics of these wards are given below.

Civil Lines (*Ward No. 23*) is inhabited by high 'class', rich people, mostly belonging to administrative and judicial

TABLE 17.6 : **Livestock population in five selected wards of Allahabad city, 1983**

Particulars	23 Civil Lines	24 Rajapur	28 Mumford-ganj	30 Shiv Kuti	34 Tagore Town
	Ward Number				
Cattle	440	1,206	569	886	391
Buffaloes	650	1,217	673	1,793	581
Sheep	—	38	70	153	—
Goats	436	755	526	1,214	230
Horses	—	23	12	60	40
Donkeys	9	13	5	4	—
Camels	—	3	—	4	—
Pigs	58	289	443	510	1,200
Poultry	305	682	51	960	248
(i) Hens	247	580	36	809	188
(ii) Cocks	58	102	15	151	60
Ducks	5	74	—	—	46

Source : Fieldwork conducted in April, 1983.

jobs and avocations. This ward has one of the best grid iron pattern of roads with big bungalows with wide open space on both sides.

Rajapur (*Ward No. 24*) has a mixed population mostly inhabited by economically backward classes of people. Once it was a village. It was incorporated in the city in the early part of this century and urbanised.

Mumfordganj (*Ward No. 28*) is inhabited mostly by the middle class people specially advocates and those working in the offices. A few backward and scheduled caste families are also there.

Shiv Kuti (*Ward No. 30*) is a part of the flood plain of the river Ganga. It provides good pastoral ground. Several new buildings have come up recently. It is inhabited by a mixed group of backward, scheduled caste and a middle class people.

Tagore Town (*Ward No. 34*) is inhabited by middle class families besides some backward and scheduled class people.

The animals are kept by the backward and scheduled caste communities. Yadavas, for example, by and large, earn their livelihood by selling milk which is in great demand in the city. In fact the population of cows and buffaloes in the city appears to be well distributed perhaps in response to the supply and demand for milk. All the five wards in question have large number of buffaloes. They outnumber cows because the former give more milk than the latter. Population of the city depends entirely on these cattle for milk because there is no organised dairy.[4]

Donkeys are used by the washermen for carrying the dirty as well as washed clothes. Horses are used for pulling cart. Pigs are kept by scheduled caste people. They play a dual role ; they supply meat, and also function as scavengers. The latter function is quite important in a city where sanitation is poor. People appear to have accepted livestock population as part of their life and they are seldom bothered about it.[5] There are only occasional protests but such protests are very mild because they know that they have to finally depend on buffaloes and cows for milk, and poultry for eggs and so on. On an average, a Yadav family possesses ten buffaloes and five cows ; but there are some who have 20 or even 40 heads ! Their social standing depends on the number of buffaloes they have. There are a few individuals of other communities also who keep cows or buffaloes but the milk is used only for family consumption.

Changes in livestock population

Table 17.6 shows only the ward-wise distribution of animals. If this distribution pattern is taken as proxy for the rest of the city, it will be very closely related to the distribution of human population. The greater the density of human population the greater the number of livestock. Thus, as the population of various wards increased the number of animals also increased but rather faster. Table 17.7 shows the temporal trend in the livestock population over a period of five years (1978-83) for two wards —Shiv Kuti and Civil Lines.

TABLE 17.7 : **Livestock population in two wards of Allahabad city during 1978-83**

Particulars	Ward Number: 30 Shiv Kuti 1978	30 Shiv Kuti 1983	23 Civil Lines 1978	23 Civil Lines 1983
Cattle	496	886 (78.6)*	216	440 (103.7)
Buffaloes	720	1,793 (149.0)	297	650 (118.9)
Sheep	509	153 (−69.9)	—	—
Goats	747	1,214 (62.5)	106	436 (311.3)
Horses	50	60 (20.0)	—	—
Donkeys	4	4 (0.0)	—	9
Camels	9	4 (−45.6)	—	—
Pigs	524	510 (−2.6)	35	58 (65.7)
Poultry	1,034	960 (—7.2)	270	305 (13.0)

Source : Fieldwork conducted in April, 1983

*Figures in parentheses show per cent variation.

One can observe from this table that the population of cattle and buffaloes has increased rather at an alarming rate. In both the wards the rate of growth of·buffalo population is greater than that of any other livestock, that is, 149 per cent in Shiv Kuti and by 119 per cent in Rajapur. Apparently this is in response to increasing demand for milk in the city. While the population of sheep is declining, the number of goats has shown an upward trend (62% in Shiv Kuti and about 311 per cent in Civil Lines). Horses and donkeys have also registered increase in their number. The population of pigs has increased by 66 per cent in Civil Lines. Of late, pigs' meat has become very popular and is in great demand.

Table 17.8 shows the decadal change in livestock population in Allahabad city during 1944-82. This table confirms the trend revealed from Table 17.7. There has been a phenomenal increase in the number of pigs (1,221.0%) during the last decade (1972-82). The phenomenal growth in the pigs population is evident from the fact that their number has gone upfrom 614 in 1944 to about 80,000 in 1982. Similarly the number of buffaloes has touched about 60,000 in 1982 from that of 2,758 in 1944. They have thus increased more than 20 times over a period of 40 years. During the last decade itself there was about fourfold increase in the buffaloes population.

TABLE 17.8 : **Decadal variation in livestock population of Allahabad (urban), 1944-82**

Particulars	1944	1951	1961	1972	1982
Cattle	8,893	12,159	13,287	24,457	47,520
(Cows and oxen)	(—)	(36.7)*	(9.3)	(41.6)	(86.6)
Buffaloes	2,758	3,586	3,793	14,744	60.232
	(—)	(30.0)	(5.8)	(288.7)	(308.5)
Sheep	683	416	117	2,933	2,014
	(—)	(—39.1)	(—71.9)	(240.7)	(—31.4)
Goats	7,895	7,276	7.091	17,846	25,825
	(—)	(—7.8)	(—2.5)	(151.6)	(44.7)
Horses	1,583	834	—	1,594	1,820
	(—)	(—47.3)	(—)	(—)	(14.2)
Donkeys	469	621	—	1,496	2,040
	(—)	(32.4)	(—)	(—)	(36.3)
Camels	—	155	—	113	160
		(—)	(—)	(—)	(41.6)
Pigs	614	519	—	6,073	80,259
	(—)	(—15.5)	(—)	(—)	(1,221.0)
Poultry	10,744	5,661	8,004	40,459	N.A.
(Fowls)	(—)	(—47.3)	(41 4)	(405.5)	

Source : District Census Handbook 1951-61, District Sta tistical Bulletin, 1977 and Municipal and Revenue Records. Figures for 1982 are based on projection. It may be mentioned here that official figures are not very reliable. Since keeping livestock in urban areas is against the rule, owners do not give correct information. This also keeps them avoid taxes.

*Figures in parentheses show per cent growth or decline of livestock.

It is also important to mention that cows have not grown in the same proportion. Yet they have also increased significantly from 8,890 in 1944 to 47,500 in 1982. Similar is the case with donkeys, horses and camels. This phenomenal increase in the population of livestock in Allahabad as is also the case with other urban centres in India reveals a reverse trend. In fact, the livestock population should increase more in rural areas than in the urban areas, but the growth in rural areas in Allahabad is far too low as compared to urban areas (Table 17.9).

TABLE 17.9 : **Per cent variation in livestock in Allahabad (rural) 1944-82**

Particulars	1944	1951	1961	1972	1982
Cattle	—	12.6	15.3	10.7	–10.8
Buffaloes	—	10.8	4.4	15.2	16.9
Sheep	—	–5.8	28.9	30.6	–10.0
Goats	—	0.4	5.3	27.7	–34.6
Horses	—	6.2	—	—	12.8
Donkeys	—	6.0	—	—	20.8
Camels	—	16.6	—	—	20.3
Pigs	—	–34.2	—	—	–6.0
Poultry	—	–13.5	–11.2	108.8	–10.0

Source : District Census Handbook, 1951-61, District Statistical Bulletin.

From the above table it may be observed that growth rate of livestock population in the villages of Allahabad district is very moderate. In fact, during the last decade the population of cattle, sheep, goats, pigs, except that of camels, donkeys and buffaloes has registered a declining trend. Is it because the animals have moved to the city ?

Urban livestock ecology

Livestock industry is space consuming. Where do the animals live in the city ? What do they eat and what arrangements are there for excreta disposal ?

Animals are found in all parts of the city but in recent years new colonies with wide roads and open space for parks etc. are the main attractions. They are kept on the footpaths on both sides of the road. It all starts with a few buffaloes. Then the lines of buffaloes go on lengthening until the end of the road is reached. Gradually, almost the whole road is covered with buffaloes and their excreta and only pedestrians and cyclists can negotiate them.

Apart from the roadsides, all vacant spaces are invaded and occupied whether it is a children's park or public land for buildings. In these vacant spaces, one finds buffaloes in knee-deep filth consisting of their excreta and mud. The whole place starts stinking and nobody other than the owners of the animals will live there. The vacant houses are then rented to lower and lower middle class people who cannot afford to pay high rent for better environment.

The cow and buffalo excreta are converted into cakes for cooking : part of it is left to be cleared by the municipal authorities. In these localities, not all the houses have latrines. Children defecate on the roadside. The pigs come from time to time to clean up the defecation but in turn release their own excreta which remains there to be spread around by men and animals passing that way.

In due course, the situation becomes so difficult from the point of view of sanitation that the municipal authorities also neglect these areas and allow the filth to accumulate. These areas thus become breeding grounds for all types of diseases and, of course, crime.

The animals are stall fed but they cannot be tied on the road during the day ! After milching, they are then carried to pasture land outside the city or to the open spaces within the city. They are taken in herds of 25-50 through the main thoroughfares of the city releasing dung as they go making the roads dirty. Then they are brought back in the afternoon the same way. When the herds of buffaloes move, traffic stops for, unlike the cows, buffaloes are not sensitive to fast moving traffic.

Allahabad is known for large colonial bungalows with huge open spaces. Each of the bungalows has a number of servants quarters. Many of the old occupants of the bungalows have left the city. Even otherwise, only a feudal society can maintain such buildings. The descendants and relatives of the servants, however, continued to occupy the out houses. Now they have laid claim over the property they occupy. Almost all of them have a few head of cattle. This has led to the entry of livestock even in the posh areas of the city. The ruralisation process is thus spreading fast in every nook and corner of city at an alarming rate.

CONCLUSIONS

This paper has presented the preliminary results of a recent study on ruralisation of Indian cities. The paper is based on limited data pertaining to livestock. More detailed information about the employment and income generated by the livestock industry within the urban areas is yet to come. What is apparent, however, even from this limited data, is that most of the urban centres in backward regions of India have reached a point of not only economic stagnation but also of economic regression. They are functionally too weak to support even the natural increase in their population. The rural periphery is, however, so poor that the immigration to the city continues unabated. This has led to a curious situation of migrants moving to the city with their rural functions giving rise to ruralised urban landscapes in many of these cities. These rural functions are no more confined to small enclaves : they now cover large areas and are well distributed all over the city.

This new phenomenon needs to be studied in greater depth and its implications for urban and rural development policies noted. This paper is only a preliminary attempt in this direction.

REFERENCES

1. The population declined during 1901-1921 due to epidemics especially due to plague, but it was a milestone in the town planning period because on the recommendation of the Plague Commission, Town Planning Improvement Trust came into existence in different and important towns of India. The Improvement Trust of Allahabad

came into existence in 1921. It became a Municipal Corporation in 1960.

2. Misra, H.N. (1982). Rural roots of urban poor: Study of an informal sector of Indian city, in R.P. Misra (ed.), *Rural Development: National Policies and Experiences*, Vol. 4 (Singapore: Maruzen Asia) pp. 215-229.
3. Brush, John E. (1962). The morphology of the Indian cities, in Roy Turner (ed.), *India's Urban Future*, (Berkeley and Los Angeles: University of California Press), pp. 57-70. Brush observes, "...the typical urban centre contains congested old section adjacent to which may be found carefully planned and often spacious sections dating from the British period. The urban morphology of India, thus presents either conflict or blending of indigenous features and the hybridized European features."
4. There was one cooperative dairy about ten years ago which survived only for a very short duration. There are a few private dairies running on co-operative basis but they are not very successful.
5. The municipal laws and civic administration are ineffective in dealing with the stray cattle problem. Besides, the owners especially the Yadavas are very militant people and do not abide by the rules.

18

SPATIAL PERSPECTIVES IN SOCIO-ECONOMIC DEVELOPMENT FROM NATIONAL AND REGIONAL ANGLES

L.S. BHAT

INTRODUCTION

THE INDIAN economy in its structure and spatial organisation continues to exhibit sharp contrasts and duality as between rural and urban components. The contribution of agriculture and allied activities to GNP is as high as 50 per cent. The rural population constitutes 76.3 per cent of the total and the population distribution shows a tendency towards infilling in areas of relatively high density which also correspond with those of agriculturally rich Indo-Gangetic plains and the coastal area girdling the peninsular plateau. The urban population of 156.19 million (1981) constituting 23.7 per cent of the total has a major share in the GNP contributed by the secondary and the tertiary sector. In terms of the areal spread, and the number of locations, the dichotomy between rural and urban components is more pronounced. For example, the number of villages in the country is approximately 0.5 million in contrast to 3,245 urban centres and the spatial distribution of urban centres and large villages (of over 5,000) also follows closely the overall pattern of population distribution (*Appendix I-II*).

Due to sharp regional contrasts in physical environment, resource base, socio-cultural and economic development, mobility of population across the length and breadth of the country is severelly restricted. Added to these, the language factor, low level of literacy and skills create artificial barriers across the states and thus limit large scale outmigration. This is corroborated by the data on migration which reveal that intra-district and inter-district (within a state) migration are more dominant than inter-State migration. Thus, even in A.D. 2000 when the total population of the country reaches around 950 million to one billion the spatial pattern of distribution would not seem to be basically different from the pattern which has persistently remained peripheral to the peninsular plateau. Likewise, the pattern of urbanisation would seem to follow the overall pattern of distribution with only 30 to 35 per cent of the total population living in urban centres and the large cities (above 50,000 population) claiming a major proportion of urban population, secondary and tertiary activities and infrastructure facilities and amenities.[1]

The resource structure, its spatial pattern of distribution and the emergence of major centres of economic activities based on these resources also bring out marked dualism, but with a difference which is so important in the formulation of a development strategy appropriate to different regions of the country. Prior to Independence, the pattern of spatial organisation of the economy was centred on major metropolitan centres/ports, capitals of British provinces and princely states. The agricultural areas were hinterlands for export of primary products or in processing, leaving traditional markets with very little of agro-industrial and related activities. In a sense, the traditional market towns (weekly and regular markets) functioned effectively as a link to the non-agricultural occupation based on households and cottage industries spread widely in the rural areas. This was evident in the relatively high proportion of rural workers engaged in non-primary activities prior to 1951. With the improvements the in transport network and technological development, the rural areas and small towns lost these activities under competition from urban high cost technology. Likewise, major industrial centres located close to their resource-base have developed 'islands' of economic development surrounded by extensive areas of under, development.

Agricultural development has, however, brought about marked changes in spatial organisation of the economy. Within agriculturally rich areas such as in the Punjab-Haryana-Western Uttar Pradesh, the Krishna-Godavari delta in Andhra Pradesh and the Coimbatore plateau in Tamil Nadu, intra-regional differences in land productivity and overall levels of development have lessened. Growing inter-regional disparities among agricultural regions of varied cropping pattern have, however, added new problems relating to the provision of even the minimum levels of social facilities and amenities and employment particularly to the small and marginal farmers, landless labourers, rural artisans, including fisherman, backward classes and tribal communities.[2] The city system marked by urban agglomerations and location along major transport routes has accentuated the dichotomy as between rural and urban into (i) rural-rural, (ii) rural-urban, and (iii) urban-urban types with varied pattern of spatial organisation in different regions of the country. Keeping in view the regional contrasts and the rural-urban dichotomies this paper attempts to identify the aspects of the spatial structure and processes of development from a national and regional angle.

SPATIAL DIMENSIONS OF THE DEVELOPMENT PROCESSES AND DISPARITIES

Regional framework

The importance of spatial dimension in economic development of a large country of contrasts is recognised in the long term perspectives of development (1960-75) and in successive Five-Year Plans.[3] Yet the planning process has incorporated very little by way of preparation, implementation and evaluation of the plan programmes on a regional consideration. This is because in the federal set-up the Centre-State relations in terms of financial allocations and fixing physical targets of production are assumed to have in built regional dimensions while they do not exist. The states do not have the adequate expertise and administrative machinery to translate the sectoral allocations on the basis of the peculiar conditions and needs of different sub-regions, districts, development blocks or clusters of villages. A critical survey of various attempts in identifying regional dimensions in planning in response to the stated objectives of resource use and minimisation of regional disparities in development reveals that these studies do not lack either in the availability of data or in techniques of measuring association among attributes and regional differentiation in the extent of association among attributes (regionalisation). For example, a regional framework for resource development was identified as early as in 1961 in which major regions and sub-regions were identified stating characteristics of the resources and norms of regional specialisation in long-term development (Rao, V.L.S.P. and Bhat, L.S. 1962).[4]

These exercises were a beginning in planning spatial organisation of the national economy which in the long run would lead to the formation of macro-economic regions bringing together economic activities of regions with complementary resources. Since districts (about 500 in number, and each state has districts of varied number and size) were used as the administrative units for boundary delineation, such an exercise provided for disaggregation of national targets on a regional basis and aggregation or translation into development plans at different levels of regions without conflicting with the recognition of linguistic state as 'region' in the Five-Year Plans. In the same exercises in regionalisation, alternative approaches to formation of economic regions were suggested.[5] One was to group linguistic states on the basis of contiguity, complementarity of resources and economic inter-dependence e.g. South India comprising four states of Kerala, Tamil Nadu, Karnataka and Andhra Pradesh and Western India with Maharashtra and Gujarat. The second alternative was to regroup resource regions and sub-regions which cut across state boundaries. The latter was more realistic for spatial analysis from the regional angle while grouping the states in an effort to the formation of macro-economic regions would have hastened the process of economic integration of states and decentralisation of decision making at different regional levels. A development framework prepared from the regional angle would by necessity have two dimensions : one is a *spatial framework* (a landuse and settlement plan) and the other *sectoral* (or departmental in the form of aggregates of physical targets and financial allocations or requirements).

Disparities in the levels of development are inbuilt in the very process of development if one takes the view of the complex relationship expressed in the formula structure↔process ↔ stage in a spatial setting which is present in all the three interrelated components. This view is recognised in the plan strategy which states that "ending disparities in levels of development between regions" (horizontal disparity) is as important as that of disparities among different sections of the society (vertical).[6] That the reduction in horizontal disparities in itself would contribute significantly to the reduction of vertical disparities has not found expression in the development strategy (*Appendix III*).[7] The developed regions such as the Punjab-Haryana though unique serve to corroborate the view that in regions whose norm of specialisation is rooted in the initial stage to agriculture, the intra-regional disparities are minimum. However, because of the sharp contrasts in agriculture and pattern of settlements, spatial organisation of the economy of these regions would be different and hence the need for preparing the development plans from the regional angle.

MEASUREMENT OF REGIONAL DISPARITIES IN LEVELS OF DEVELOPMENT

Studies in measuring the variations in levels of development among regions have not lagged behind in following up the stated objectives in the Third Five-Year Plan. Here again district has been the unit of data compilation, while choice of indicators (mostly reflective) shows wide variations because of the paucity and inadequacy of indicators such as per capita income and expenditure on essential items of consumption. The techniques of analyasis have ranged from working out composite rank scores to the application of factor analysis. Since quite a few under-developed regions are extensive and show marked affinity to environment, resource base and initial level of development the boundaries of least developed regions such as semi-arid Rajasthan, south-east resource region (comprising forestry and mineral resources of low density and tribal population), parts of the Decean plateau and the north-eastern hill region coincide with those of the regions and sub-regions having homogeneity in formal characteristics. Composite ranking of states has also been attempted by the Special Committee of the Planning Commission and social scientists.[8] Because of large size and heterogeneity the ranking of states in the development scale does not provide a true picture of development/under-development [9]

Development problems of regions

Studies from the regional angle carried out in parts of the hill regions of Uttar Pradesh (U.P.) and southern Bihar (Chota Nagpur) have brought out the issues relating to the spatial process of under-development that are partly due to inherent topographical and resource characteristics and the regional setting. In the case of the Uttar Pradesh Hill Region,[10] the situation itself limits inter-regional and intra-regional transactions with international boundary to the north and east and by being a hill-forest sub-region of the State of Uttar Pradesh. Most parts of U.P. being an alluvial plain with agriculture dominating, the hill region has always been a periphery to the main areas of economic organisation. Low land productivity, small sized settlements (84 per cent of the population living in villages of less than 500 persons—1971) and inaccessibility (over 63 per cent of the villages being more than 5 kms away from the main roads) have meant low levels of social and economic interactions within the region. The main road passing through this region connecting the religious and tourist centres to the north has helped to siphon the primary products from forest and migrant population. Sex ratio is highly imbalanced (1,160 females per 1,000 males) and the region is under the vicious circle of low land productivity ↔ out-migration ↔ poverty with money order remittance of small amounts supplementing the income. The sample survey of households and places of out-migrants revealed that migration is male selective and the migrants are from relatively accessible areas to the towns or contact points within the region. The literacy, as in some hill areas in north-east India, is relatively high (41%) and so is female work participation. Priority for development in this area is one of increasing the land productivity in the available cultivated area to stabilise the population and of increasing the social facilities and amenities as part of the minimum needs programme on the basis of individual settlements and not on a selective basis as part of the programme for removal of poverty.[11]

The study of Chota Nagpur plateau region (southern Bihar) of basic and heavy industries, a paradoxically underdeveloped region, has different characteristics.[12] While the region has similar constraints as the U.P. hill region in respect of topography, low population density, scattered, small sized settlements and inaccessibility, its economic interaction with far flung urban-industrial centres such as Calcutta and Bombay is foot-lose so far as regional economy is concerned. The tribal population, with its low level of consumption and subsistence agriculture does not have the intermediary levels of infrastructure, and activities for vertical integration with the large industries. There are, however, internal variations in spatial organisation of the regional economy centred on market places which are either small towns or large villages. Attributes of development—demographic, social facilities and amenities, economic activities and infrastructure such as electricity and roads measured by composite score of settlements and the development blocks brought out nine typologies of development/underdevelopment within the Chota Nagpur region which itself stands out as underdeveloped at the national and state levels (*Appendix-IV*). At the State level there are some empirical evidences which bring out the fact that the State of Bihar with all its rich agricultural area and major industries is less developed partly because the three sub-regions e.g. the tarai (foothills) the central plain and the southern plateau (Chota Nagpur) do not have economic interaction.

The Western Ghats region is yet another example to illustrate the need for evolving a development strategy from the regional angle in which three area levels are involved for the purposes of formulating a development strategy within the stated goal of integrating (i) environment in the economic development strategy, (ii) physical and human resource aimed at specialisation in the cultivation of plantation crops with adequate green leaf manure, (iii) water use plan for a large number of tributary valleys instead of giant hydro-power projects, and (iv) afforestation of the much denuded hill tops and valleys. Here the narrow valleys of tributary streams with small sized settlements dotted along the stream constitutes the basic and the lowest areal unit for planning and groups of such valleys and interfluves constitute the second higher order units as brought out in the study of the Karnataka portion of the Western Ghats.[13] At the third order regional level, the sub-regions of the Western Ghats forming parts of Maharashtra, Goa, Karnataka, Kerala and Tamil Nadu have a dual role, first in relation to the economy of the concerned states and the optimum utilisation of forest, and more particularly of the resources of the Western Ghats region from the point of view of the economic development of all the southern states and Maharashtra. The alternative strategy for water use for energy and irrigation has conflicting interests in the regional and national development strategies. Submergence of forest areas and human settlements displacing the population from its cultural and environmental ties has created new problems including problems of health and disease as in parts of the north and south Kanara districts in Karnataka. Spatial organisation of the economy itself has been external to the Western Ghats region on account of its configuration and topographical conditions and the region is now undergoing changes in its functional economic orientation with the commissioning of the West Coast Highway and increased accessibility without corresponding economic inputs and social facilities and amenities.[14]

STRATEGY FOR REMOVAL OF POVERTY

Development of different regions and removal of poverty are complementary in the Indian context. The strategy of minimisation of regional disparities stated in the national objectives are far from the geographical realities of poverty and development prevalent in different regions. This is apparent in the twin strategies adopted in the Five-Year Plans. Firstly, the strategy for planning from the grass-roots level is still in its experimental stage starting from the identification of growth centres in the context of preparing District Development Plans during 1970s to the emphasis on block level planning in the Fifth Five Year Plan.[15] The stress on identifying less developed areas during the 1960s had to be modified because of the complexity in the measurement of under-development areas and the political overtones attached to it. Instead, special assistances were provided for a variety of characteristics of under-development e.g. hill areas, tribal areas, command areas, drought-prone areas and so on, often with overlapping of some schemes within the same area and implemented by different agencies in isolation.

The Sixth Five Year Plan made a significant departure by adopting the household as the unit for direct attacks on poverty while generally stating the need for preparing block and district plans according to the resources and needs of different areas. The new strategy aims at amelioration of the conditions of nearly 75 million of the poorest people by choosing six households per annum from each village in all the 5,000 development blocks or half a million villages in the country. Since these programmes are so varied and highly localised there is no scope for its continuity in the context of the development of the block as a whole. A few exercises in evaluation of these schemes, just begun, bring out two basic drawbacks in their implementation. Firstly, the choice of the six poorest families in a village is not devoid of subjectivity. Secondly, the programmes, such as of providing a milch animal on a subsidised basis, do not have built-in linkages with other activities or land uses that support them, for example, raising fodder locally, collection of milk and access to chilling centres. Whereas such linkages are spontaneous in agriculturally rich areas as in Punjab-Haryana as is evident by the emergence of small dairy centres, and farmers finding it economical to raise grass in lands which hitherto were used for wheat.

The Priority for preparing block plans and integrating poverty programmes within them can not be over emphasised.

REGIONAL FRAMEWORK FOR INTEGRATING ENVIRONMENT, RESOURCE DEVELOPMENT AND AMELIORATION OF POVERTY

The twin issues relating to the minimisation of regional disparities in development and amelioration of poverty discussed above lead us to the search for regional typologies according to environment, resources, human settlements and spatial organisation.

Contiguous and extensive areas of relatively high levels of development with minimum variations within

These conform to the agriculturally rich areas as in the Punjab-Haryana, western U.P., Krishna-Godavari delta, Mysore plateau, Coimbatore plateau and Cauvery delta. In some of these areas the market towns show a distinct regularity both in spatial distribution and heirarchic pattern of development. A fourtier hierarchic pattern of nodality with Delhi at the apex is noticed in the studies of Karnal district in Haryana,[16] and Muzaffarnagar district in Uttar Pradesh.[17] There are definite indications of the trickle down effect of development impulses with small rural settlements of even 1500-2500 population range acquiring infrastructure facilities and amenities. There are, however, certain differences in the location and growth of newly created small market centres. Such markets located along radial routes converging on major market towns/regional economic centres, as, for instance, Karnal and Panipat, are bypassed by the farmers who prefer to avail of multiple transactions and officiency in clearing the market arrivals of commodities and spot payment of cash. Such rural-rural difference in the pattern of integration of different size range of settlements might not be a short term anomaly. These anomalies are also noticed as between the inner delta area of the Krishna-Godavari tract and the contact zone between the delta and the foothills in Andhra Pradesh. The latter has emerged as a distinct agro-industrial region with Guntur-Vijayawada-Rajahmundry as the foci of development.

Linear pattern of development with limited spread effect on the surrounding area

Such patterns of development have been characteristic of the coastal plains in Karnataka and Kerala. In these areas, the pattern of distribution of population itself is elongated conforming to older sand dunes and coastal plains. The transport route also follows the same pattern. While the contiguity of settlements makes it difficult to differentiate the boundaries of different settlements, the differences in levels of functions of different towns and large villages (of over 10,000 persons as in Kerala) is clearly seen in the breaks in the intensity of movement of people and various commercial, social and infrastructural facilities and amenities distributed along the highway passing through the settlement. In a sense the spatial organisation of the economy is linear-oriented to market villages and towns. Such a spatial organisation of the economy has rendered the mid-land and foothill areas of Kerala and the parts of coastal Karnataka peripheral areas while their norms of specialisation are rooted to grassland, livestock and plantation economy.

Contiguous and extensive areas of low levels of development with weakly developed market towns and/or administrative centres

These areas are mostly agricultural areas with very little diversification and intensity of cropping. They extend over parts of the Deccan peninsula comprising western Madhya Pradesh, Vidarbha (part of Maharashtra), the central and northern Bihar, West Bengal, Orissa and the Northern *maidan* (plateau) area of Karnataka. The spatial organisation of the economy has inherently a weak agricultural base and does not show a well-developed hierarchic system of nodal centres.

Areas of low levels of development with inherently low agricultural productivity

These include drought-prone areas like the Rayalaseema and the semi-arid red soil plateau of Telangana in Andhra Pradesh and the semi-arid Rajasthan. In these areas urbanisation itself is low with an imbalance in urban hierarchy. Likewise the larger urban industrial centres like Hyderabad-Secunderabad have very little influence on the surrounding area.

Hilly and tribal areas of extensively low levels of development with low levels of resource use

These areas exhibit a variety of spatial patterns of organisation depending on the situation as discussed in the foregoing pages, for example, hill areas of Uttar Pradesh, Chota Nagpur in Bihar, eastern Madhya Pradesh and Orissa comprising roughly the extensive tribal belt of the country.

Areas of low levels of development due to the impact of the metropolitan cities

The impact of the metropolitan city on the surrounding region has brought about marked imbalance in the growth of a system of market town production centres. The impact of Bombay on the Konkan region of Maharashtra has been so widespread that the traditional market towns which used to be the centres of collection and distribution of consumer goods have declined. This is an instance to illustrate how increased accessibility without corresponding economic inputs accentuates the externality of space relations. Such examples are numerous along the axis connecting the major cities of India.

Concluding remarks

The regional framework for rapid social and economic change in the Indian context does not appear to be kaleidoscopic as is made out in studies on regionalisation. It is true that in the search for homogeneity in environment, resources and pattern of organisation, regions can be differentiated at different scales of areas using village, tahsil (or development block), and the district and all these are relevant depending on the socio-economic attributes that are important at different scales of areas. Excessive reliance on macro- and meso level aggregated indices of development and using them for regional differentiation according to levels of development do not by and large serve to answer the causes for widening or narrowing of regional disparities as the spatial structure and processes of organisation are not built into these studies. Secondly, the development plans prepared from below are presently no more than aggregation of schemes formulated by different organisations without using spatial approach to planning. The planning machinery equipped with expertise and infrastructure does not exist even at the district level but for a small nucleus provided by the Planning Commission. In the political set-up the regional parties coming to power in some of the states, people expressing strong regional sentiments and aspirations and the desire to have a second look at the Centre-state relations in a much wider framework point to the need for *regional* approach to planning and priority for decentralised decision-making with the goal of economic integration of states and their sub-regions.

REFERENCES

1. The projections made by various organisations vary between these percentages. International Bank for Reconstruction and Development has worked out the percentage for urban population at 32.2 per cent in A.D. 2000. *The Task Ahead for the Cities of the Developing Countries*, I.B.R.D., 1975, Washington, D.C.
2. This is evident in the strategy for amelioration of the rural poor. Government of India, Planning Commission, *Sixth Five Year Plan* (1980-1985)
3. Government of India, Planning Commission. *Third Five Year Plan* (1961-66).
4. The above statement is re-emphasised by Hanumantha Rao, C.H. (1979) "Poverty and Development—Characteristics of Less Developed Regions in India", *Economic and Political Weekly*, Special Number, August 1979.
5. Bhat, L.S. (1971). Ref. Bibliography. During the same period the Town and Country Planning Organisation also worked out a similar exercise but suggested a larger number of economic regions and sub-regions.
6. Government of India, Planning Commission, *First Five Year Plan*, 1951-56.
7. This view is also corroborated by the observations of C.H. Hanumantha Rao in *Poverty and Development—Characteristics of Less Developed Regions in India*, Ibid., p. 1807.
8. Government of India (1968). *Report of the Committee on Identification of Backward Regions* (Pande Committee Report). According to this

Report, Uttar Pradesh, Bihar and Andhra Pradesh are among the industrially less developed states

9. Aggregation of small areas into large and heterogeneous units eliminates important characteristics and problems of under-development and hence the problem of under development itself from the National and State angles.
10. Dobhal, G.L. (1983) *Development of the Hill Areas—A Case Study of Pauri Garhwal District.* Ph. D. Dissertation (unpublished).
11. The strategy for removal of poverty during the Sixth Five Year Plan (1980-85) aims to cover annually 600 families per Development Block (there being 5000 blocks with 100 villages in each) i.e. it is proposed to cover 75 million people by identifying 6 poorest families annually in each village and providing them employment opportunities on the basis of locally selected schemes. An amount of Rs. 1500 crores is provided in the Sixth Five Year Plan and in addition Rs. 3000 crores are being provided through institutional finance. Estimate of income generated annually is expected to be of the order of Rs. 3000 crores (Sixth Five Year Plan : 1980, pp. 51-53).
12. Sinha, Savita (1983). *Regional Structure, Patterns and Processes of Development—A Case Study of Chota Nagpur Region,* Ph.D. Dissertation, Bhagalpur University (unpublished).
13. The northern part is suited to forestry and mineral utilisation, the central part for forests, spice gardens and paddy cultivation. The southern part is suitable for forestry—Plantations (coffee, cardamom and fruits and paddy).
14. Bhat, L.S. (1982). North Kanara Districts—A study in spatial organisation of the District and Regional Economy. (mimeo). This observation is corroborated by another study, Deshpande, C.D. *et. al.* (1980) : *Impact of a Metropolitan City on the Surrounding Region—A Case Study of South Colaba (Konkan) in Maharashtra,* Concept Publishing Company, New Delhi.
15. Development Block is smaller than a district and comprises 100 villages with approximately 1,00,000 population. Government of India, Planning Commission (1978) : *Report of the Working Group on Block Level Planning* also refer Banerji, S. *et. al.* (1973) *Hierarchic Location Analysis for Integrated Area Planning : Experiences of the Pilot Research Project in Growth Centres,* India, paper presented at the Regional Science Congress, Vienna, 1973.
16. Bhat, L.S. (1976) Refer Bibliography
17. Rao, V.L.S.P. (1976) Refer Bibliography.

SELECTED REFERENCES

Alam, M. *et al.* (ed.) (1982). *Settlement System of India*, Oxford & IBH Publishing Co. New Delhi.

Bhat, L.S. (1976). *Micro-level Planning*, A case study of Karnal Area, Haryana, India, K.B. Publications, New Delhi.

Bhat, L.S. (1972). *Regional Planning in India*, Statistical Publishing Society, Calcutta.

Bhat, L.S. *et al.* (1975). *India Essays on World Urbanization*, Ronald Jones (ed.), George Philip and Son Ltd., London.

Bhat, L.S. *et al.* (ed.) (1982). *Regional Inequalities in India*. Society for the Study of Regional Disparities, New Delhi.

Bronger, D. (1978). Central Place System Regional Planning and Development in Developing Countries : Case of India, R.L. Singh *et al.* (ed.) *Transformation of Rural Habitat in Indian Perspective*, National Geographical Society of India Research Publication No. 19.

Desai, P.B. *et al.* (ed.) (1969). *Regional Perspective of Industrial and Urban Growth*, Macmillan and Co., Bombay.

Dasgupta, Biplab (1971). Socio-economic classification of districts—A statistical approach, *Economic and Political Weekly*, Vol. VI, No. 33, 1763-1774.

Deshpande, C.D. (1980). *Impact of a Metropolitan City on the Surrounding Region: A study of South Colaba, Maharashtra*, Concept Publishing Company, New Delhi.

Dobhal, G.L. (1981). "Money-order Remittances and Outmigration—A Case Study of Pauri Garhwal District" *Annals of the National Association of Geographers, India*, Vol. I, No. 2, 75-79.

Folke Steen (1968). "Central Place Systems and Spatial Interactions in Nilgiris and Coorg (S. India)", *21st I.G.U. Congress*, India, Collected Papers, Denmark.

Ginsburg, Norton (ed.) 1960. *Geography and Economic Development*, University of Chicago, Department of Geography, Research Series.

Government of India, Ministry of Works and Housing (1977). *Report of the Task Force on Planning and Development of Small and Medium Towns and Cities*, Vols. I and II.

Kundu, A. and Raza, Moonis (1982). *Indian Economy : The Regional Dimension*, Jawaharlal Nehru University, New Delhi.

Hemlata, Rao (1982). *Regional Disparities—Dimensions and Typology of Backwardness*, Institute for Social and Economic Change, Bangalore, Mimeo.

Misra, R.P. *et al.* (1969). *Regional Planning Concepts, Techniques, Policies and Case-Studies*, Prasaranga, 1969.

Misra, R.P. *et al.* (1980). *Multi-level Planning and Integrated Rural Development in India*, Heritage Publishers, New Delhi.

Mitra, A. (1965). *Levels of Regional Development*, Census of India 1961, Vol. I, Part A (i) and (ii).

Nair, K.R.G. (1982). *Regional Experiences in a Developing Economy*, Wiley Eastern, New Delhi.

National Council of Applied Economic Research, New Delhi (1965). *Market Towns and Spatial Development in India*, New Delhi.

Pal, M.N. (1963). A method of regional analysis of economic development with special reference to South India, *Journal of Regional Science*, Vol. 5, No. 1, pp. 41-58.

Ramachandran, H. (1980). *Village Clusters and Rural Development*, Concept Publishing Company, New Delhi.

Rao, V.L.S.P. and Bhat L.S. (1962). "A Regional Framework for Resource Development", *Bombay Geograghical Magazine* Vol X. pp. 35-50.

Rao, V.L.S.P. (1963). *Regional Planning*, Indian Statistical Institute, Calcutta.

Rao, V.L.S.P. (1976). *Development Strategy for an Agricultural Region*, Institute of Development Studies, University of Mysore, Mysore.

Registrar General and Census Commission of India : Census of India, 1981. Provisional Population Totals—Rural-Urban Distribution, Paper 2 of 1981.

Sen, Lalit *et al.* (1971). *Planning Rural Growth Centres for Integrated Area Development : A Study in Miryalguda Taluka*, National Institute of Community Development, Hyderabad.

Whyte, R.O. (1982). *The Spatial Geography of Rural Economies*, Oxford University Press, Bombay.

APPENDIX—I

POPULATION DISTRIBUTION (RURAL AND URBAN) IN SETTLEMENTS OF DIFFERENT SIZE CLASSES, 1971 CENSUS

RURAL

Population size	No. of settle-ments	% to tolal No. of settle-ments	Popula-tion (mn)	% to total rural population
Less than 200	1,50,072	26.06	15.25	3.47
200—500	1,68,561	29.27	56.43	12.89
500—1000	1,32,990	23.09	94.41	21.50
1000—2000	81,973	14.23	113.60	25.78
2000—5000	36,005	6.25	104.55	23.82
5000—10000	7,974	1.58	32.72	7.45
Over—10000	1,358	0.24	22.33	5.08
Total	5,75,933	100.00	439.05	100.00

URBAN

Size class	Population	No. of settle-ments	% to total No. of settle-ments	Popula-tion (mn)	% to total urban population
VI	Less than 5,000	290	9.30	0.89	0.82
V	5,000—10,000	820	26.29	6.20	5.68
IV	10,000—20,000	987	31.64	13.96	12.80
III	20,000—50,000	652	20.90	19.95	18.29
II	50,000—1,00,000	219	7.02	14.71	13.48
I	1,00,000—5,00,000	134	4.30	26.49	24.28
	5,00,000—10,00,000	9	0.29	6.12	5.61
	Over—10,00,000	8	0.26	20.77	19.04
	Total	3,119	100.00	109.09	100.00

Source : Rao, V.L.S.P. (1983). *Urbanisation in India—Spatial Dimensions*, Concept Publishing Company, New Delhi.

APPENDIX—II

URBAN POPULATION IN DIFFERENT SIZE CLASS OF TOWNS—1971 AND 1981

	Size class	No. of UAS/ towns		% to total number of UAS/towns		Urban population (mn)	% of urban population to total		Growth rate	
		1981	1971	1981	1971	1981	1971	1981	1961-71	1971-81
VI	Less than 5,000	230	150	7.09	5.94	7,86,483	0.45	0.50	−19.70	+60.74
V	5,000—10,000	742	641	22.86	25.32	56,41,505	04.57	3.61	−10.38	+15.44
IV	10,000—20,000	1,048	847	32.30	33.46	1,48,62,211	11.20	9.52	+18.85	+24.03
III	20,000—50,000	739	570	22.77	22.52	2,24,13,463	16.32	14.35	+29.27	+28.41
II	50,000—1,00,000	270	178	8.32	7.03	1,81,91,847	11.24	11.65	+40.94	+51.22
I	1,00,000 & above	216	145	6.66	5.73	9,42,92,998	56.21	60.37	+52.67	+56.82
	Total	3,245	2,531	100.00	100.00	15,61,88,507	100.00	100.00	+37.91	+46.02

Source : Census of India 1981. *Provisional Population Totals.*

APPENDIX-III

INDIA : PATTERNS OF REGIONAL AND URBAN DEVELOPMENT-SELECTED EXAMPLES

Region	Districts	INDEX VALUES					PERCENTAGE INCOME ORIGINATING FROM		
		Over all development	Agri. development	Development of secondary activities	Development of tertiary activities	Urban development	Primary	Secondary	Tertiary
(1)	(2)	(3)	(4)	(5)	(6)	(7)	(8)	(9)	(10)
1. Relatively developed regions									
Coimbatore Plateau	Coimbatore	1.67H	1.01H	1.60VH	1.30VH	1.37VH	36.95	24.48	38.57
	Salem	1.25M	1.02M	1.25H	1.09H	1.19H	49.65	20.42	29.93
	Madurai	1.26H	1.08H	1.38VH	1.31VH	1.40VH	39.91	18.59	41.51
Krishna-Godavari	E. Godavari	1.13H	1.15VH	1.29H	1.16H	1.28H	55.42	12.99	31.59
	W. Godavari	1.26H	1.16VH	1.29H	1.11H	1.22H	59.07	9.58	31.35
	Krishna	1.22H	1.09H	1.22H	1.14H	1.35VH	57.63	7.31	35.06
	Guntur	1.24H	1.12H	1.45VH	1.04M	1.27H	61.03	14.90	24.37
Punjab Plains	Ludhiana	1.80VH	1.19VH	1.42VH	1.31VH	1.50VH	32.56	14.26	53.18
	Patiala	1.38H	1.14H	1.19H	1.19H	1.26H	46.26	7.86	45.88
	Jullunder	1.81VH	1.17VH	1.48VH	1.30VH	1.42VH	31.05	23.25	45.70

APPENDIX III (Contd.)

1	2	3	4	5	6	7	8	9	10
2. Low level of development and low resource potential Telangana	Adilabad	0.88L	0.85ML	1.02VH	1.00M	0.89M	37.43	23.66	38.91
	Nizamabad	1.09M	1.04M	1.34VH	0.95M	1.12M	44.94	27.28	27.28
	Karimagar	1.10M	0.94L	1.14H	1.12H	0.79L	32.75	29.25	38.00
	Warangal	1.02M	0.93L	1.25H	1.06M	1.13H	39.98	17.73	42.29
	Khamam	0.87L	0.95M	0.92L	0.92M	1.01M	47.22	21.28	31.50
	Medak	0.83L	0.94L	0.84L	0.99M	0.79L	44.56	19.47	35.97
	Hyderabad	1.93VH	0.91L	1.34VH	1.69VH	1.69VH	11.76	8.10	80.14
	Nalgonda	091M	0.91L	0.86L	1.00M	0.96M	41.79	17.87	40.34
	Mahbubnagar	0.86L	0.90L	0.90M	0.91L	0.80L	49.86	15.87	34.27
B. Marathwada	Aurangabad	0.77L	0.88L	0.94M	0.90L	1.08M	55.44	11.71	32.95
	Parbhani	0.77L	0.87L	0.87L	0.96M	0.89M	54.63	8.70	39.58
	Bhir	0.71L	0.88L	0.69L	0.84L	0.83L	61.13	6.96	31.91
	Osmanabad	0.72L	0.85L	0.63VL	1.11H	0.83L	30.16	3.30	66.54
	Nanded	0.77L	0.82L	0.99M	0:99M	1.04M	43.99	15.75	40.26

APPENDIX III (Contd.)

1	2	3	4	5	6	7	8	9	10
3. Low level development and high resource potential South-East Resource Region	Bilaspur	0.75L	0.97M	0.85L	0.73VL	0.93M	71.60	6.26	22.14
	Raipur	0.81L	0.97M	0.98M	0.84L	1.05M	65.08	8.97	25.95
	Surguda	0.65L	0.92L	0.55VL	0.54VL	0.57VL	70.89	12.78	16.33
	Sidhi	0.58VL	0.93VL	0.59VL	0.50VL	0.34VL	82.33	4.96	12 71
	Baster	0.63L	0.91L	0.54VL	0.47VL	0.52VL	83.27	5.80	10.93
	Korapur	0.74L	0.80L	0.57L	0.94VL	0.66M	58.24	6.07	35.69

EH-Extremely High ; VH-Very High ; H-High-M-Medium ; L-Low : VL-Very Low ; The District Income Data are only illustrative although income orginating approach has its limitations.

19

THE UNEXPLORED REALM

C. D. DESHPANDE

THE parameters of a science are no less influenced by the surrounding social and economic conditions in which its scholars find themselves. This is particularly true of social sciences including geography. Geography has passed through its various stages of description, determinism and possibilism, natural regions, geomorphology, quantitative revolution and spatial analysis. Environmental and ecological awareness, perception and behavioural aspects are the aspects now engaging our attention. A new consciousness is that of a commitment by geographers to studying society ; its ills and disorders. This is increasingly regarded as the core of social geography (Jones and Eyler, 1977). The present essay tries to outline this commitment in the Indian context and indicates how Indian geography could be usefully oriented to these new directions in order to serve Indian society.

HUMAN REACTION OR RESPONSE TO THE EXTERNAL WORLD

This theme was elaborated by G.P. Marsh as early as in 1842, but found a vigorous expression in the writings of Ratzel and his followers including Griffith Taylor ; but though rebutted to a large extent by Vidal de la Blache and his followers, it does persist, albeit indirectly, in modern writings (Martin, 1951). The issue of determinism vs possibilism was largely rendered, and rightly, irrelevant by spatial analysts. But the new emphasis on environmentalism stems from ecology and has gathered good momentum with the growing concern about conservation of natural resources and preventing environmental pollution. All these aspects, however, have one thing in common. They examine the impact on human society of factors that are external to man.

HUMAN RESPONSE TO THE 'WORLD WITHIN' AND ITS EXTERNAL EXPRESSION

The history of thesocial sciences shows that 'Man' was taken as a normative human being. In 'Economics', he was the person endowed with wisdom, forethought and the necessary purchasing power to satisfy 'consumer's demand'. In sociology, he was a member of a family, a clan and a tribe, obeying the dictates of tradition, taboo, religion and magic, till Malinowski hinted at the need for studying his mind and mental processes. In geography, he was the person who inexorably followed the commands of nature, one who had many choices before him and that he, in his wisdom, would make the right choice.

It is due to the developments in psychology and medical science that the need to understand human response to his environment, in terms of his perception, field of information, and his behaviour influenced by emotions and beliefs, has been increasingly felt. Alexis Carrel (1948) emphasises how great is the complexity of the human mind and how we are only at the fringe of understanding it. Sauer referred to the cultural aspects which interpose between man and nature. More recently, Yi-Fu Tuan (1974) has elaborated on the strong emotional ties between man and his home. Not surprisingly the study of the individual and his 'private geography' has found its scholars like P. Gould (1972) and others. Indeed, the gains in psychology and social anthropology have contributed to this development. The essential argument in developing this aspect of geography seems to be that it is as important to study the individual's perception and behavioural patterns within his social and natural setting as to study the external environment in which he dwells as an individual, and as a member of his ethnic, religious and social groups (Kirk 1975,

Brookfield, 1969). This today is largely an 'unexplored realm'' in geography, and how we may benefit ourselves by such studies will be elaborated with Indian examples in the remaining part of this essay.

REGIONAL PLANNING

Let us start with the problem of regional socio-economic development especially of the backward regions. Under the national planning policy for the last thirty-odd years, it would not be too much to say that the planning inputs have been largely decided by the policy makers, administrators and local leaders, without understanding the scale of priorities as envisaged by the regional communities, nor of their outlook and ability to accept innovations placed at their disposal under planned expenditure. The result is there for all to see ! The targets achieved usually relate to the amounts spent, and sometimes in terms of the physical facilities created, but very rarely, indeed, in terms of benefits actually derived by the people.

This is particularly true of development of tribal areas where not only the ecological balance is upset in the name of commercial forestry, but facilities like irrigation, domestic water supply, and improved agricultural methods are created, but not effectively used by the tribal community. The newly constructed roads drain away local natural resources, and possibly encourage outmigration. The new formal schools are either not used or used partially without any enlightenment on the part of the tribal community as regards use of innovations. Extension services and demonstration centres operate with a limited success. The tribal remains internally bound with his limited perception, and inhibitions arising out of religion and tradition, and externally with his incapability in making use of the planned facilities, due to poverty and ignorance of new technology as well as lack of economic facilities by way of credit and co-operative organisation. It is time that we study perception and behavioural responses of the tribals in relation to their environment that stands threatened with technological changes and exploitation by external agencies like the forest contractor or the mining lease holder. It is also known that irrigation development in a tribal area does not get the required response from the community and supply of water for domestic use leads to neglect of its proper drainage and gives rise to malaria and water-borne diseases.

URBAN ENVIRONMENTS

Although so much attention is given by our scholars to urban morphology, urban systems and the like, how is it that very little attention is paid to the urban dweller and how he perceives his environment ? It would appear that the business tycoon, the middle class office-goer and the slum dweller have each their own view of the city. The businessman's perception of his city possibly consists of his residence, his office, bank, five star hotels, airport and the roads connecting them ! The office-goer's urban world is conditioned by his commuting to his office. A recent study of slum dwellers shows that a worker in Bombay knows nothing beyond his route to the place of work ! (Chavan 1976). All this, however, needs to be studied in detail.

Again, it is well known that one can be very lonely in a populated city ! In a social sense, the self-contained flat while providing modern services and facilities, heavily operates against one's social instinct, and bars social contacts with the neighbours. The Indian city, thus, seems to have lost its civic virtue, and has become an intensely crowded and ill-organised space in which the only concern of the dweller is to earn and survive ! This tendency, observed in general, needs testing.

True to the Indian tradition, the Indian city has its caste and community as well as immigrants' residential enclaves. Such traditional enclaves, 'the pols' in Gujarat cities, assured the needs of the residents very well. Regardless of their economic status, the residents lived as a neighbourhood group with a sense of social and economic obligations to each other (Deshpande, 1953 ; Desai, A., 1980). It is necessary to see how these 'pols' are functioning today, what replaces them, and how the new socio-cultural zones of a city evolve and function.

For such studies, the behavioural patterns and processes of the residents need to be analysed.

Excessive crowding and its impact on the urban mind is yet another fascinating aspect of study as has been shown by studies of the Ahmedabad city (Kulkarni, K.M. 1983). It seems to be a world in itself! Human basic needs, social and religious customs, cultural activities, means of earning, all these are heavily conditioned by cramped space. A study of these aspects might lead to a better formulation of policies for providing municipal services and social facilities like educational institutions, and more importantly for imparting more effective civic education.

URBAN VILLAGES

Indian metropolitan expansions have engulfed the neighbouring villages. These 'urban villages' have been studied from the point of view of their form and structure (Sundaram, K.V. 1977). Not only is it necessary to study the growing urban impact on the economic conditions of the residents, but also the social distress and the enforced migration that it entails and the anguish it generates. A recent study of the villages in Delhi Metropolitan Region aims at measuring the quality of environment as perceived by the residents (Ouiji, 1983). But more needs to be done as regards the choice given to the residents and their ability to make use of the aid given for their being uprooted. Indeed, newly settled areas like Madangir in the Delhi Region await such studies.

RURAL ENVIRONMENTS

While our scholars are engaged in studying settlement hierarchies *ā la* Christaller, settlement morphology and the like, a thought needs to be given to our village residents in relation to their social environment. It is not enough to identify caste enclaves and their composition. More important it is to know why they occur and persist, and what changes, if any, are taking place. Two recent studies of rehabilitated villages whose old sites were submerged under a dam of the Godavari and another on the Krishna (Sawant, 1983) reveal that the village residents do not like to have their new houses in 'mixed' localities, and they prefer to exchange their allotted plots so as to form a caste enclave ! This shows that the process of social integration, desirable as it is as a long-term policy, cannot be hastened, and the attitudes of the residents need to be studied in order to identify the right policies and strategies. This consideration applies with even greater force in the case of the Harijan enclaves, and the problems in amicable sharing of facilities like the village water supply.

Again, while canal irrigation brings immense benefits to command areas and makes the rich farmer richer as is seen in Maharashtra, the plight of the small farmer, especially in the non-command areas, needs to be studied. Steeped in his poverty and insecurity, his attitudes and abilities need to be understood before we can give him the right relief and encouragement. The same is the case of the landless labourer who has perhaps lost his small holding and become footlose in his search for employment.

DROUGHT PRONE ENVIRONMENTS

We have become all too familiar with governmental measures to aid drought prone areas : employment relief, loans, cheaper foodgrains and famine relief measures like well construction and road building programmes. But it is well known that these measures—at heavy government cost—have a limited and temporary efficacy. It is more than necessary to understand the perception, attitudes and behavioural patterns of the drought stricken individual, his family, and his village community, so as to understand better his scale of preferences in relief and the level of his acceptance of skills that are necessary in innovative agriculture and newer industrial processes.

Drought prone areas are notorious for the distress exodus of their residents. Though the Indian census data and governmental statistics are available to us, we do not know enough of the outlook of the drought prone individual, his problems and his ability to withstand harsh conditions. Even as things stand today, he is a lone warrior fighting his battle with aature, hoping for the rains, trying to hold his home ground as best as he can, and often giving up the game and migrating to cities in the search for survival (Ambhire D., 1981). Indeed, we hardly know anything of this poignant aspect of man-environment relationship. It is necessary to know how the drought prone individual can be served better, by way of offering, for example, 'holding sites' in selected towns where he can be temporarily accommodated, given relief and training, in order to rehabilitate him after the famine in his home village.

TRIBAL AND NOMADIC ENVIRONMENTS

We have already referred to the need for understanding the tribal mind and attitudes when we referred to the problems in regional, social and economic development. A recent study of the Warli tribe in Maharashtra (Thane district) revealed that it is neither the ignorance nor the inhibitions on the part of the tribal folk that prevent them from using HYV rice variety, but sheer poverty and unimaginative administration that were responsible (Janani, 1981). The tribal opposition to commercial forestry sometimes bursts out in the form of agitations like the *Chipko* movement in the Garhwal-Kumaon Himalayas, and sometimes in the form of violence as in south Bihar and eastern Madhya Pradesh. To understand the problems involved in this type of man-environmental relationship is eminently a geographer's task.

Nomadic communities in all the regions of India play an important role in regional economy. They are the migrating expression of a type of man-environment relationship. Yet, sufficient attention has not been paid to them by geographers. The studies that we have stem from anthropologists and sociologists as well as to the scholarly administrators of the earlier generation. How fascinating and how necessary it is to study nomadic communities from a geographical point of view, can be cited with one example. The Third Five-Year Plan proposals of the state government to settle Lamanis and other nomadic tribes of Maharashtra by providing housing including a primary school soon came to naught as the tribes in pursuit of their traditional occupation just could not settle down. Even now it is necessary to understand their needs, the scale of preferences in development and their capabilities. This is equally true of the shepherd communities which in Maharashtra are significant in number. So are they in Gujarat, Rajasthan and the Himalayas ; for that matter in almost every region of India. It is really difficult to understand why geographers are not attracted to such problems in man-environmental relations.

THE CHANGE IN THE POINT OF VIEW

It will be seen from the foregoing examples that it is necessary to change our point of view. In addition to study the geographical phenomena in the traditional manner, that is, studying 'spatial patterns and processes' as an observer or scanner with a checker board, we might change our point of view and begin studying man-environment relationship of the man concerned. This is the fascinating and unexplored realm which geographers might explore with benefit to the nation and as a rewarding exercise for themselves.

REFERENCES

Ambhire, D. (1981). "Drought Prone Areas and the Human Response", (Ph.D. thesis in preparation).

Brookfield, H.C. (1969). "On the environment as perceived", *Progress in Geography*, Vol. II.

Chavan, V. (1976). "Kalina", M. Phil. thesis, Bombay University.

Carrel, Alexis (1948). "Man the Unknown", Pelican.

Desai, A. (1980). "Ahmedabad City", ICSSR Project.

Deshpande, C.D. (1953). "Visnagar", *Geographical Outlook.*

Gould, P. (1972). *Mental Maps,* Penguin.

ICSSR Review Report (1971, *et al.*). Geography, Sociology, Anthropology, Psychology.

Janani, R. (1983). "The Warli Tribes and Agricultural Innovations", (Ph.D. thesis in preparation).

Jones, E. and Eyles J. (1977). *An Introduction to Social Geography*, Oxford University Press.

Kirk, W. (1971). "Problems of Geography" in *Readings in Social Geography*, Oxford University Press.

Kulkarni, K.M. (1983). *Crowding in Ahmedabad City*, ICSSR Project.

Sundaram, K.V. (1977). *Urban and Regional Planning in India*, Vikas.

INDEX